样板房设计新法(下)

NEW
DESIGN
METHODS
OF SHOW
FLAT

设计献策

BRAINSTORMING
ON DESIGNS
AND SCHEMES

深圳视界文化传播有限公司 编
Shenzhen Design Vision Cultural Dissemination Co., Ltd

中国林业出版社
China Forestry Publishing House

序言 PREFACE

A good show flat is the conciseness of culture and space, which concentrates the designer's highest inspiration of life. From the gorgeousness at the beginning of the cooperation, it turns into maturity and wisdom. If we say that the birth of a show flat is just like the production of a film, then I am the producer, while Daguan design is the director who is good at controlling. The studio is similar to a battlefield, where there is spark with smoke of gunpowder.

——Yuxiang Zhang
General Manager of Landscape and Interior Design Center of Greenland Group

好的样板房是文化和空间的凝炼，集中了设计师对于生活的最高感悟。从合作之初的惊艳而后愈发走向成熟和睿智，如果说一套样板房的诞生如酝酿一部电影，我是制片人，大观·自成则是个非常有控制力的导演，片场如同战场，火花伴着硝烟。

—— 绿地集团景观与室内设计中心 总经理

张雨翔

I have been in Shanghai for almost 15 years, who have almost witnessed the development of real estates within these years. The market of show flat has been mature since the attempts and finally become "inundant". Thus, designs learn and facsimile from each other to aggravate the "homogenization phenomenon" in the market, and similar appearances occur in the designs of show flat. However, the consumers are more rational than before with the changing purchasing psychology, then design comes back to the individual level and the essence again. Design should be a way to lead a new concept of life and advocate styles.

Designers express business elements with design languages, and combine personal experiences to present a value, which are the methods to break homogenized competitions and reduce replaceable risks. The advent of customization will be changed on the designers. With the rise of the east, eastern languages become the obbligato trend of design. Even though we have not experienced the industrial revolution, we still have the mission to let the west view eastern design and culture, while the truth is that they are already expecting to see our transformation.

A design should start from the demands of the real consumers nowadays to take consider of the true match between living functions and culture, for example, whether the combination of the spatial design is reasonable on the living habits, traditional behavior patterns, spatial generatrix and so on. For instance, in the Jinhua Villa with neo-oriental style we designed, we considered the relevance and extensibility of the interior and exterior environments with traditional cultural preference on the design of the basement, where the art furnishing of pine trees inside together with the shadows made the partition between interior and exterior space vague.

The re-invention of eastern culture is to look back constantly—searching for the source and excavating the stories about the techniques in the renaissance.

Since I have worked in this industry for more than 20 years, I still persist "Good Design" in my interior projects. The "Good Design" in my mind is delicate and impeccable. For example, our team consider the "inclusiveness" of the space as

CHALLENGE·BREAKTHROUGH·INCLUSION

挑战·突破·包容

much as possible in the design without excessive decorations to restrict the imagination of the client, we have to meet the demands of all family members with individualities in different spaces. While this is the so-called "a top-grade residence cultivates personal integrity with inclusiveness"! We attempt to break the definition of "beautiful appearance" on show flat. Since the refined decoration progress is relatively complex, we try our best to make an incorporate technological process from the exterior structure to the interior decoration. The interior layout which seems like insignificant determines whether a design is good or not. It is not a metaphysical appearance. Constant and patient coordination of the design team is needed to combine the interior configuration with the architecture closely and perfectly.

Besides hard decoration, each of the soft decorations is like a poem. I create the projects from the view of sense organs such as visual sense, auditory sense and olfactory sense, to motivate the emotional experiences of people who step in and then make them feel each delicate detail at home through the sense of touch. Experiencing as if one were standing inside of it is the starting point of a design, for each product is built on premise of "home" and the orientation of design is making the expectations of home and dream to the extreme.

"Quality comes from a good design, while only a good design could create high-quality goods." We hope these designs with refined decoration could influence the market gradually, which could be popularized as a demonstration model.

Daguan design association
Jacky Lien

我来到上海近乎15年,也算见证了这15年地产的发展。看着样板房市场在经历摸索的过程中成熟,最后到"泛滥"。于是,设计之间的互相学习临摹,市场的"同质化现象"加重,样板房的设计趋近相似的外观。然而由于消费者越加的理性,购买心理在不断的改变。设计又回归到个性的层面,回归设计本质,设计原本就是引领新的生活概念,倡导风格。

设计师将商业的元素用设计语言表达,结合个人自身的经历去表述同一个价值,才是突破同质化竞争的手段,减少了可替代性的风险。定制化时代的来临,这一切都会在设计师身上得到转变。当东方的崛起,东方的语言变成不可缺少的设计趋势。即使我们没有经历工业革命的这个缝隙,但是我们依旧有着让西方重新看待东方设计文化的使命,而事实是西方也正在期待看到我们的转变。

设计应从当下现实消费者群体的需求出发,考虑其居住功能配置与文化的真正匹配。如生活的习惯,传统行为模式与空间动线,空间设计的结合是否合理。例如我们设计的锦华别墅新东方风格的一户,在地下室空间设计上,考虑了传统文化喜好的内外环境关联性和延伸性,内置的松树园林艺术陈设和光影交错的设计将室内外的划分变得模糊。

东方文化的再创造,就是不断地回头审视——寻根,挖掘那些复兴工艺的事儿。

从业20多年以来,我在室内设计中所一贯坚持的就是 "Good Design" 优良设计。我认为的"优良设计"就是要做到精致且无可挑剔。例如,在项目设计上,我们团队尽可能的去考虑空间的"包容性"——不去做过多装饰性的设计从而限制客户的想象,在不同空间里,对家庭角色的全面照顾,以及产生的功能需求,甚至个性上发挥的余地。所谓,上品居所,独善包容!试图去突破那种"看起来很美"的样板房的定义。因为精装修施工环节相对复杂,我们尽力要做到从外部结构到室内装修几乎一体化的流程。看似微不足道的内在布局,都决定了一个作品的好坏,绝非形而上学的表面功夫。需要设计团队耐心地不断协调,将室内配置与建筑的关系完美地紧密结合。

硬装之外的软装,每个作品就像是一首诗,这种对于多少的斟酌绝不多一点或少一点,场所精神的发挥。我从感观的角度去创作,从视觉、听觉、嗅觉来激发步入者的情感体验,再从触感去感受家中每个精妙的细节。身临其境的体会是设计的出发点,每一件产品都是以"家"为前提打造,设计目的就是让人们对家以及梦想的期待完全被诱发出来。

"精品来自于优良设计,一个好的设计才能打造出精品。" 我们希望所做的精装修的设计,慢慢去影响着市场,更是作为一个示范性典范被推广。

大观·自成国际空间设计 连自成

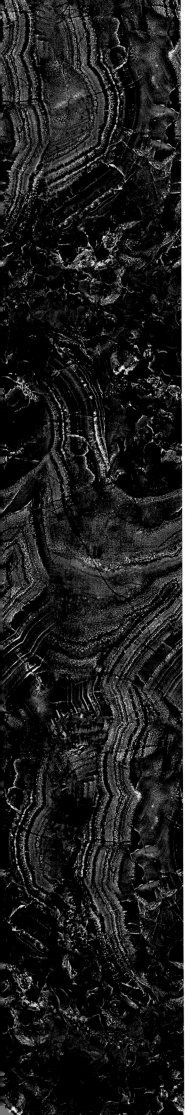

目录
CONTENTS

AMERICAN STYLE
美式风格

- **008** THE REVERIE OF BIRDS
  飞鸟的遐想
- **016** AMERICAN MANOR
  美式庄园
- **032** INHERITING AMERICAN SPIRIT
  传承美式精神
- **040** A FANTASY WORLD
  幻彩世界
- **052** PLEASANT HOURS
  惬意时光
- **058** HEMES MEET WILD PEOPLE
  HEMES 遇见狂野分子
- **066** A SOLEMN AND ELEGANT HOME
  庄重典雅之家
- **074** THE STORIES OF TIME
  光阴的故事
- **084** SEARCHING FOR THE SPRING
  清景寻春
- **092** ENJOY THE TIMES
  静享漫时光

NEO-CHINESE STYLE
新中式风格

- **102** AN ELEGANT STATE
  素面雅境
- **112** QUIET LUXURY WITH ELEGANT FLAVOR
  低奢雅韵
- **122** CLEAR ZEN FLAVOR WITH FRESH DREAMS
  素禅·清梦
- **130** LIVING IN THE IDEAL STATE
  栖居桃花源
- **140** LITERARY FRAGRANCE IN THE MAGNIFICENT MANSION
  书香致远 屋华天然

150	TIME IN THE CORNER OF THE CITY 城隅时光线	*180*	THE FLOWING FRAGRANCE 暗香浮动
162	CHINESE RESIDENCE WITH ZEN FLAVOR 中式禅居	*192*	SCULPTING IN TIME 雕刻时光
170	THE FRESH ZEN 清雅禅音	*200*	FEEL QUIETLY 静悟

MODERN LUXURIOUS STYLE
现代奢华风格

216	THE PURITY 繁华落尽见真淳	*286*	A LIFESTYLE DESTINATION 寻找生活的归宿
224	THE SUNBATHE 沐阳	*292*	TIME COLLECTION 时光收藏
236	MIRROR·CLEANNESS 镜·净	*296*	FADDISH CITY 时尚都市
248	THE FUTURE HOME 未来之家	*300*	ARTFUL LIFE 艺术生活
256	THE ELEGANT LIFE 典雅生活	*308*	METROPOLITAN 精品大都会
264	A RHAPSODY IN ILLUSION 幻景狂想曲	*316*	KEY C WITH FASHION 时尚C调
276	FRAGRANCE THROUGH TIMES 时光倾吐芳华		

SOUTHEAST ASIAN STYLE
东南亚风格

324	A VACATION IN SOUTHEAST ASIA 度假东南亚	*348*	A PURE REGION 净域
332	LOVE IN SOUTHEAST ASIA 情定东南亚	*358*	A CHANSON IN GREEN 绿茵香颂
338	A SOUTH ASIAN RESIDENCE WITH ZEN FLAVOR 南亚禅境		

AMERICAN STYLE / 美式风格
NEO-CHINESE STYLE / 新中式风格
MODERN LUXURIOUS STYLE / 现代奢华风格
SOUTHEAST ASIAN STYLE / 东南亚风格

AMERICAN STYLE
美式风格

THE REVERIE OF BIRDS
飞鸟的遐想

◇ DESIGN CONCEPT ◇

In a winter, desserts collocate with the warm sunshine, you can sit outdoors to stare the birds and clouds, then you will achieve a good mood. You could see Erhai with the smoke from kitchen chimneys and the boat yesterday faintly, while this space enjoys the sceneries of Dali without the disturbances of the visitors, where people could enjoy the leisurely time quietly. Designers clear up the space relations in the semi-basement to bring the gentle sunlight into the house so that people will forget their sorrows. You could plan an ideal afternoon, walk leisurely in the vegetable market, select food materials for relatives or friends and prepare for a gorgeous meal.

设计公司：品辰装饰工程设计有限公司

硬装设计师：庞一飞、袁毅

软装设计师：张婧、夏婷婷

项目地点：云南大理

项目面积：180平方米

主要材料：做旧实木地板、硅藻泥、水曲柳木饰面、麻布布艺等

◇ 设计理念 ◇

冬日暖阳，甜点搭配日光，坐在户外坐席，看着飞鸟白云。光是这样呆呆地望着心情就会很好。隐隐约约可以看到不远处的炊烟和昨日泛舟的洱海，这样的空间纵享大理的所有，没有观光客的叨扰，能让人静静品味闲暇时光。设计师将半地下室的空间关系重新梳理，目的是让可以看见的柔和日光潜入室内，让人忘忧。策划一个理想的下午，与清风一起散步，逛逛菜市场，亲自为亲人或者朋友，挑选食材，准备丰盛的一餐。

Here, you could find the imaginary world which is inenarrable in life and brew up many fresh ideas, making the power of inspiration accumulate continuously. Custom designed Persia silk rug, hand-made sheepskin lamp and warm lights indoor make people have the eager to stay at home for a whole day. Coming into Dali for many times, the expectation of freshness has been enhanced. It seems that you must absorb something different here. The pureness and plainness as well as the abundant living sense of old times could give the inhabitants a long aftertaste.

在这里可以发现生活中难以描述的想象世界，酝酿出许多鲜活的灵感，让创意能量不断累积。定制的波斯地毯、羊皮手工灯、室内的暖色光线，让人想窝在家里一整天。多少次到大理，新鲜感的期望值，已被它不断提升，感觉总要吸收些许与众不同，这里的纯粹、朴素及丰富的老时光生活感让居住者足以回味数年。

AMERICAN MANOR
美式庄园

◆ DESIGN CONCEPT ◆

As soon as entering the house, you will feel the beauty of manor with natural sense. American style is used as the design principle to give the owner the most suitable plan in order to play an elegant and graceful music. Opening the door, you will enjoy the warm texture as if coming into an American manor in your dream, where the warm sense of home is exuded in the laying of gentle and pristine materials, fabric sofas and the European carved fireplace, as well as releases the upscale taste of the hostess. In addition, the main wall with different tone is adopted with dark blue and white, outlined with white boards, making it appear a unified and rich angle to look up and presenting the purest temperature in the space. Entering the living room with ocean flavor, gray-blue tone and small retro decorations are selected to embellish the style, where you could walk into a green home with agile steps, see the vast scenery of countryside, breathe as you like and smell the fragrance of rice and flowers. The paces of life should be so slow and pleased to experience the most natural leisure time.

设计公司：北京山禾金缘艺术设计有限公司

配饰设计师：郭红梅

助理设计师：王超、侯雪莲

项目地点：山东青岛

项目面积：331平方米

主要材料：装饰画、壁花、墙体彩绘、吊灯等

◆ 设计理念 ◆

在入室之初感受自然荟萃的庄园之美，以美式风格为设计主轴，将最适切的规划留予屋主，轻谱出典雅的曼妙乐章。敞开大门，彷佛来到梦想中的美式庄园，享受温馨的质感，在温婉与质朴的质材铺陈下、布料沙发或欧式雕花壁炉，让家盈满温暖的感觉，也释放出屋主的极致品位。此外，不同色调的空间主墙，以深蓝与米白为题，并用白色线板轻轻勾勒，使其拥有一致而丰富的仰望角度，也铺陈出空间中最纯粹的温度。来到海洋风的客厅，选以灰蓝色调与复古风格小物装点品味风格，踏着轻快的步伐，走进绿意环伺的家，放眼望去尽是辽阔的田园景致，大口呼吸，闻到的都是稻香与花香，生活步调就该如此缓慢而惬意，感受最自然的悠闲时光。

INHERITING AMERICAN SPIRIT

传承美式精神

◇ DESIGN CONCEPT ◇

This case is located in the center of Dalian territory, which is adjacent to the prosperity, where you could also enjoy the mountain scenery, lake, park, shore and forest in the garden quietly...It is just like walking at the riverside of the beautiful Hudson River. The Hudson River goes through the prosperity and splendid scenery in northeast of America, while the lakeside villa shows the highly civilization and freedom in America as well as the spirit of the locals, such as struggle, preciseness and the exploration towards freedom. Thus, the classical American style full of family inheritance spirit is used in this case, based on the powerful brown tone, presenting the magnificent manner, security and comfort, where the blue which symbolize ocean and sky and the yellow which symbolize passion and hope as well as light and warmth are used.

项目名称：亿达春田留庄样板间

设计公司：大连上乘设计有限公司

设 计 师：李佳

项目地点：辽宁大连

摄 影 师：王厅

主要材料：木饰面、仿石砖等

◇ 设计理念 ◇

本案坐落于大连几何版图的中心位置，紧邻繁华之时又可静享园区内的山景、湖泊、公园、海岸、森林……犹如漫步在风景优美的哈德逊河畔。哈德逊河贯穿了美国东北部的繁华与绝佳美景，湖畔别墅体现了美国高度的文明自由以及当地人拼搏奋斗、严谨传承与自由探索的精神。因此本案运用了富有家族传承精神的经典美式风格，以浑厚有力的棕色调为基础，不仅气魄十足，也会令人倍感安全与舒适，点缀象征海洋与天空的蓝色，以及激情与希望、光与热的黄色。

Mediterranean blue is boldly added in the children's room, while the Provencal purple makes the space more diversified, thus the color comparison of the overall space is rich and mature. On the decoration, bronze lamps and ornaments, leather furniture and the decoration in ox horn are used with the embellishment of gentle and bright flowers and fabrics, making the whole space low-key and prosperous at the same time. From the view of functions, the lines are concise, elegant and steady without losing the demeanour of this brand with a hundred years, presenting the family culture and spiritual pursuit of the owner.

儿童房更是大胆加入了地中海的蓝色，普罗旺斯的紫色，使空间色彩更加多元化，整体空间色彩对比浓郁而成熟。在装饰布置上，摒弃了繁琐与空泛，运用质感相对较强的纯铜灯具与摆件、真皮家具与牛角材质的装饰品，柔美鲜亮的花卉与布艺穿插其中，使整体空间低调与奢华共存。家具布置从功能性出发，线条简洁优雅，稳重自持不失百年品牌的风范，以此来体现房主的家族文化及精神追求。

A FANTASY WORLD
幻彩世界

◇ DESIGN CONCEPT ◇

This space could meet all of your fantasies about colors, while the extreme contrast colors will brighten your eyes when you enter here. Yellow and purple match perfectly. The background wall of the fireplace in the living room is simple-modelled yet full of individualities, collocated with the humorous decorative painting of a laughing horse, making people laugh uncontrollably. The façade of the staircase is bright yellow green which echoes with the prairie green on the wall, giving people a joyful mood. A slight Zen flavor is exuded through the elegant home atmosphere in the sitting room at the second floor, while the decorative painting is the highlight which is impressive, adding the glamour into the whole space. You could drink a cup of coffee, read books or talk with your friends and taste the fresh tea under the warm sunshine in an afternoon. The designer builds a fantasy world with the colorful dreams for the whole family.

设计公司：陆槛槛空间设计

设 计 师：陆槛槛

项目地点：浙江宁波

项目面积：300平方米

摄 影 师：刘鹰

主要材料：仿古砖、木制品、壁纸、灯具等

◇ 设计理念 ◇

这个空间满足了你对色彩的所有幻想，极致的撞色让入室者眼前一亮。黄色与紫色变得如此合拍。客厅壁炉的背景墙造型简单却个性十足，再配以诙谐的张嘴大笑的马儿装饰画，逗趣横生，让观者忍俊不禁。楼梯的立面为亮眼的黄绿色，与墙面的草原绿遥相呼应，给人带来想拾阶而上的雀跃心情。二楼起居室典雅的居家氛围中透露着淡淡的禅韵，点睛之笔实数墙上的装饰画，让人过目不忘，为整个空间增彩添色。当阳光暖照在这儿，你可以一个人抿一口咖啡看看书，也可以与朋友品着新茶畅聊一个下午。设计师为屋主打造了一个幻彩家世界，这里装载着一家人彩色的美梦。

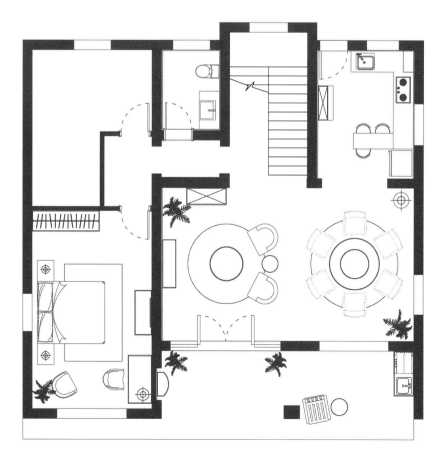

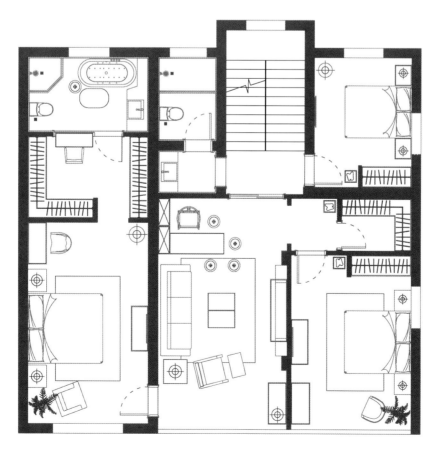

PLEASANT HOURS
惬意时光

◇ DESIGN CONCEPT ◇

This case is positioned as American style with a collection of excellent European and American design elements, which abandons the complex carvings, while the simple structure and plain colors show the living atmosphere of people coming back to nature and pursuing for comfort, leisure and old times perfectly.

Entering the living room and dinning room through the entrance garden, you will see the broadness of them. The design of the living room and dinning room is remoulded to be connected, where the arched door which is commonly seen in American country style is used to be the partition, echoed with the arched door on the wall with mellow and graceful lines as well as simple and generous structures. The living room and dinning room are dominated by light yellow, embellished with brown, blue, coffee and other supporting colors, making the space colorful and the atmosphere warm. Cotton and linen sofa is loose and soft as well as comfortable.

项目名称：佛山西樵明珠花园样板房设计

设计公司：朗昇空间设计

项目地点：广东佛山

项目面积：170平方米

主要材料：大理石、地砖、地板、涂料、墙纸等

◇ 设计理念 ◇

本案定位为美式风格设计，汇集欧美风格优秀的设计元素，摒弃其繁琐雕琢倾向，在结构造型上相对简单，色彩朴素，完美表现人们回归自然，追求舒适、休闲、怀旧的生活气息。

由入户花园进入客餐厅，便可见十分开阔的客餐厅空间。客厅与餐厅设计是经过空间改造，两厢融汇贯通的，使用美式乡村风格中常见的拱门造型进行分区，与墙面的拱门相呼应，线条圆润优美、结构简洁大方。客餐厅空间以浅黄色为主色，交错使用棕、蓝、咖啡等配色，使空间色彩丰富，氛围温馨。棉麻质沙发，宽松柔软，十分舒适。

The pristine tea table, tables, chairs and the ceiling droplight in kraft paper embody the rough natural feature in American style. While the decorative pendants on the background wall presents another kind of Roman flavor. Designers take advantage of the corner between the entrance garden and the living room, where small round table, cabinet and accessories add endless idyllic scenery into the public space, which could be used as beautiful decorations as well as a place for people to have a rest and enjoy the desserts leisurely.

古朴质感的茶几与桌椅、牛皮纸质的天花吊灯等，均展现其美式风格中粗犷的自然特色。背景墙上的装饰挂件则又呈现出另一般的罗马风情。入户花园与客厅一角亦被完美地利用起来，小圆桌、壁柜及饰品给公共空间中增添了不尽的田园风光，既作为美丽的装饰，又可以使人在此小憩，享受惬意茶点。

　　主卧室设计空间较大,以大面积浅色、蓝色为主调,以呈现浪漫、沉静的生活氛围。浅色的天花与墙面不经雕琢,将淡蓝色的床头背景包围起来。整个空间内,实木制作的床具、柜体等完美呈现出美式家私的稳重与精美。大大小小的山茶花形似随风自然飘落,点缀在窗帘、床品之上,自然且有品味。床头柜上的装饰品亦独显特色,丘比特玩偶憨态可掬,给空间增添不少乐趣。

　　主人房卫生间设计淡雅自然,洗手台柜与梳妆镜均选用实木制作,壁灯精致亮丽,装饰物别致富有情调,与外红内白、线条优美的大浴缸一起,空间内富有动感,美式风情十足。

　　次卧室与儿童房设计均沿袭使用蓝色背景,但与主人房相比,则各显特色。次卧室设计简洁大方,居家氛围浓烈,窗帘、床品及装饰柜上恣意盛开的花朵,灿烂绚丽。实木制作的床头柜、壁柜,其线条虽经简化,但依然散发着怀旧、质朴的气息。儿童房设计则通过运用色彩艳丽、活波可爱的窗帘、饰物、家私等,来创造出一个充满想象力的空间,俨然成为孩子们的生活乐园。

HEMES MEET WILD PEOPLE
HEMES遇见狂野分子

◇ DESIGN CONCEPT ◇

The designer chose the furniture and furnishings that could embody the accumulation of time and culture in this space to match up with the theme of the hard decoration, at the same time contrast colors were boldly used to neutralize the steady dark tone of the overall space, adding a little funny sense into the space. On the premise of meeting the residential function, the participation of both the designer and the owner made the living environment fit for the aesthetic needs and living demands of all the family members highly, creating the delicacy and accuracy of details to surpass the Chinese imitating of "common American style" or "common European style" on general meaning. The building of a residence contains the endeavor of the owner and the designer, as well as bears the weight of the inhabitants' hope and yearning of life.

设计公司：尚层装饰（北京）有限公司杭州分公司
设 计 师：池陈平
项目地点：浙江杭州
项目面积：335平方米
摄 影 师：叶松
主要材料：樱桃木护墙、进口壁纸、进口大理石漆、皮革、卡布奇诺石材等

◇ 设计理念 ◇

设计师在这个空间里，选择能够体现出时间与文化沉淀的家具与陈设品，以此配合硬装的主题风格，同时利用大胆撞色来中和整个空间沉稳的深色调，让空间增加些许的俏皮感。设计师与业主的共同参与使得居家环境高度贴合家庭成员的审美需求和生活需要，在满足居住机能的前提下，营造出设计细节的精致与准确，以此超越一般意义上所谓"泛美"或者"泛欧"的中国式模仿。一栋住宅的筑造，不仅倾注着主人和设计师大量的心血，同时也承载着居者对生活的期待和憧憬。

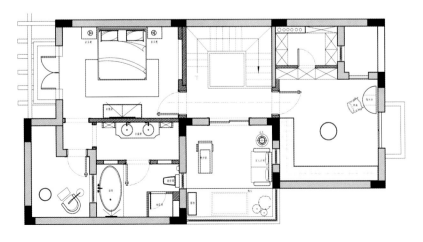

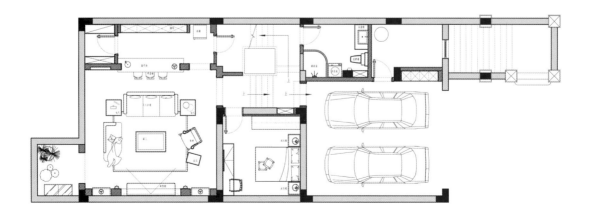

A SOLEMN AND ELEGANT HOME
庄重典雅之家

◇ DESIGN CONCEPT ◇

American style, which is luxurious yet friendly as well as bright yet comfortable, brings an intuition full of masculine power. However, handling the features of the style correctly in common American style and making the interior space filled with vivifying positive energy are not so easy. The designer uses heavy panels to decorate the façade, where classical decorative elements present the low-key yet luxurious aspect of American style, whilst dark brown also sets a steady tone in the overall home atmosphere. The managements of the ground, wall, ceiling and soft decoration are extended on the basis of such kind of a color scheme. The colors with different lightness that belong to the same tone are rich of rhythmic changes, interpreting a unified and elegant effect as a whole.

设计公司：尚层装饰（杭州）分公司
设 计 师：池陈平
项目地点：浙江杭州
项目面积：450平方米
摄 影 师：叶松
主要材料：橡木实木护墙、进口壁纸、拿铁米黄、卡布奇诺石材、皮革、进口大理石漆等

◇ 设计理念 ◇

华贵而不失亲切、明亮并且舒适，美式风格带给人们的直觉感受总是充满了阳性力量。然而，在屡见不鲜的美式风格当中，能够准确把握风格特点，让室内空间充盈着生机勃勃的正能量，却又颇有讲究，并非是一件容易的事情。设计师将厚重的护墙板装饰立面，古典的装饰元素展现出美式风格低调而奢华的一面，深棕色也为整个家庭氛围定下了稳重的基调。地面、墙面、天顶、以及软装的处理便在这样的配色方案基础上加以延伸，同色系不同明度颜色富于节奏的变化，演绎出统一、优雅的整体效果。

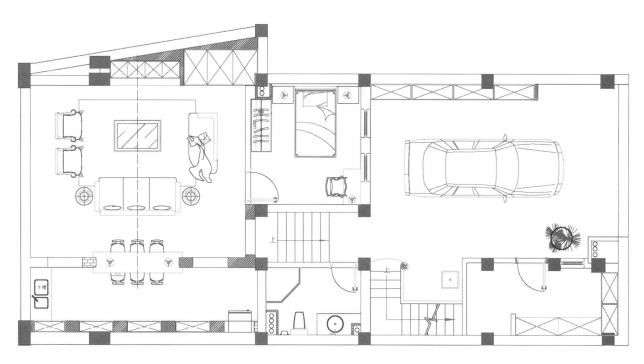

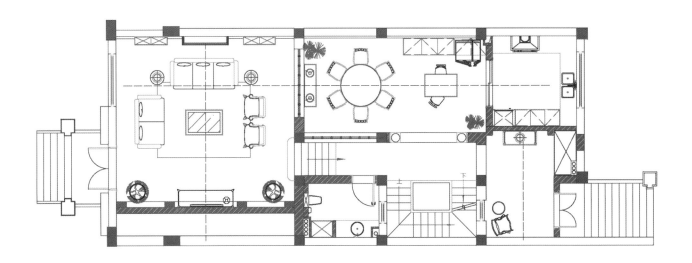

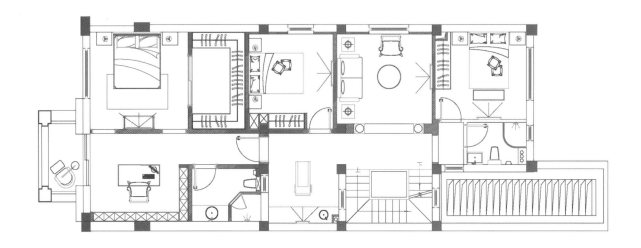

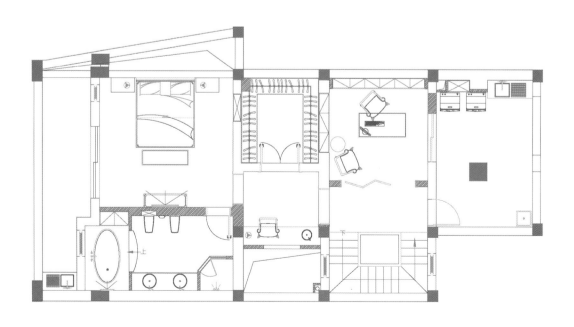

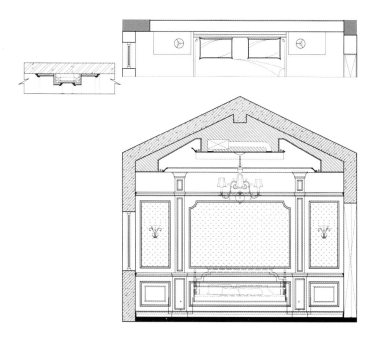
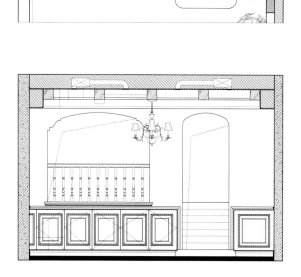

THE STORIES OF TIME
光阴的故事

◇ DESIGN CONCEPT ◇

Bleaching away the shadow of the deliberate design to restore the original flavor of life may be a higher state of design. Beauty is a kind of attitude on life, which tells the love of life from the details. In order to decorate the house, the owner and designer spent almost two years across the country to search for every destined ornament of this home. It's hard to explain the standard of the home, but the right feeling which could only be understood.

Most of the furniture in the reception area at the first floor come from CASA PAGODA in Taikang Road, Shanghai. The bleached cattle hide sofa from Italy is very solid when sitting on it. The hand-engraving solid wood decorative mirror seems that it could reflect the wonderful love stories in ancient times. Since the oak wine tub aside is always used to store wines, you could smell the intoxicating fragrance as soon as you open it. This is a collection of a boss of a soft decoration studio in Kaixuan Road for one of his French friends, which made us fondle admiringly and finally bought by us.

设计公司：贾峰云原创设计中心
设 计 师：贾峰云
软装设计：贾峰云原创设计中心 & V8陈设设计工作室
项目面积：400平方米
主要材料：大理石、仿石砖、特灰泥等

◇ 设计理念 ◇

漂白掉刻意设计的影子，还生活本来的味道，或许这是设计的更高境界。美是一种生活态度，从细节讲述着对生活的热爱。为了装饰这个家，业主和设计师花费了近两年的时间到全国各地找寻每一个和这个家有缘的物件。想要什么样的家，标准很难说的出来。感觉对了才行，而感觉，却只能意会。

一楼这个会客区的大部分家具，来自上海泰康路上的CASA PAGODA。漂白做旧的牛皮沙发来自意大利，坐上去很坚实的感觉。手工雕刻的原木装饰镜仿佛能映射出古代美丽的爱情故事。边上的橡木酒桶由于常年藏酒的缘故，一打开便会飘出醉人的酒香。这是凯旋路一家软装店的老板从法国朋友那里收来后又被我们看到，爱不释手，便买了回来。

The contrast on the depth of color in the living room and dinning room at the second floor embellishes the simplicity of life. The green vintage wall clock is especially conspicuous on the white wall, then the stories of time are counted into history. The sculpture of bird on the wall symbolizes a family of three, who could sit on the hanging chair to enjoy the leisure sufficiently under the sunshine at weekends. Dark green vintage handling is adopted on the wall in second bedroom. You could make a cup of tea, hold a good book and hear the ring on the roof to let your thought fly freely into the fantasy world in the future.

二楼的客餐厅直接的深浅对比，衬托出生活的简单。绿色做旧挂钟在白墙上格外显眼，光阴的故事在这一时刻被计入历史。墙上飞鸟雕塑象征着一家三口，周末坐在吊椅上，充分享受阳光下的惬意。次卧室墙面做了墨绿色仿古处理。泡壶茶，捧本好书，听梁上驼铃叮当，让思绪自由飞向未来奇幻世界。

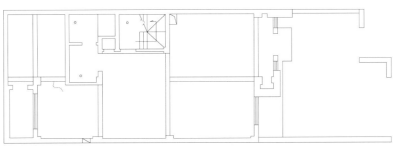
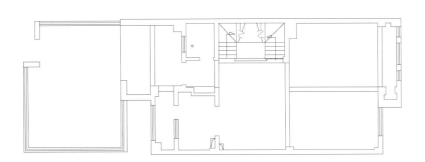

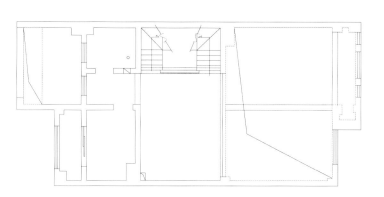
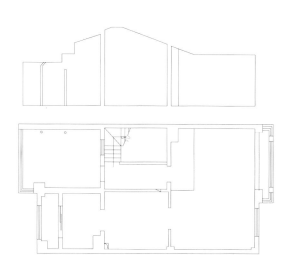

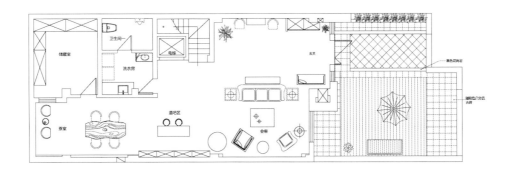

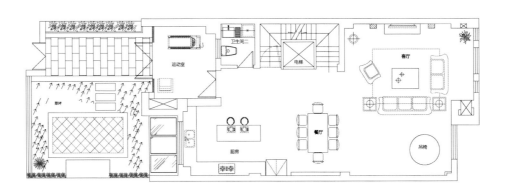

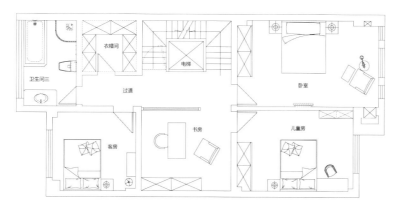

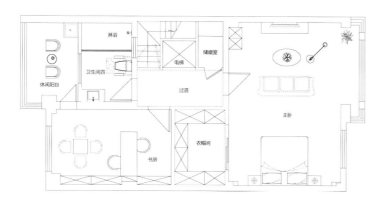

SEARCHING FOR THE SPRING
清景寻春

◆ DESIGN CONCEPT ◆

"Flower catkins are falling slowly under the setting sun." Space is just like a changeable painting which depicts infinite imaginary space through the inspiration of d creator, whilst it exudes the beautiful tone of home. Colors in the early spring decorate the whole space, where designers make the space full of the flavor of spring by the collocation of the gentle wood color and bright colors with different brightness. The puerto design of the curves softens every point of junction in the space, creating a calm and pristine idyllic flavor. The bold and distinct use on colors makes each area present the leisure atmosphere with full vitality. The floating wall surface, natural mosaics with rich texture and retro green dinning chairs combine people's emotions with the environment to create a pleased, comfortable and elegant space for the owner. From the original scheme to the selection of colors and furniture as well as the soft decorations in the late period, designers do them by themselves, presenting a spring atmosphere at last.

设计公司：嘉兴康盛装饰工程有限公司
设 计 师：蒋军华、蓝斌
项目地点：浙江嘉兴
项目面积：130平方米
主要材料：仿古砖、彩色砖、墙纸、彩色乳胶漆、实木地板、饰面板护墙、石膏线等

◆ 设计理念 ◆

"肃肃花絮晚，菲菲红素轻。"空间，彷佛是一幅变化万千的画作，借由创作者之手注入灵力，描绘出无限的想象空间，也挥洒属于家的美丽色调。初春的颜色装扮了整个空间，设计师用温润的木色配以不同明度的亮色让这个居室充满春的气息。曲线的垭口设计柔和了每个空间的连接点，营造了沉静朴质的田园风情。大胆鲜明的色彩运用，让各场域跳出充满活力的休闲氛围。行云流水的墙面，肌理丰富而自然的马赛克，复古绿的餐椅，更是将内心情感与环境融合，营造属于屋主的惬意舒雅空间。从最初的方案设计，到色彩、家居的选择，以及后期的软装配饰，设计师都是亲力亲为，最终呈现给大家的是一派春意盎然之景。

ENJOY THE TIMES
静享漫时光

◆ DESIGN CONCEPT ◆

Entering the hall, you will see the high-ceilinged space, this is the generous and majestic luxurious life that the designer attempts to present. There is a rich color comparison in the overall space, such as coffee, beige and milk yellow, as well as gold which is the highlight, rich yet not gorgeous, extruding the luxury and high quality of the whole space wonderfully. The patterned carpet consists of blue and red, which is passionate, collocated with wood and leather furniture, appearing leisurely, comfortable yet still luxurious and noble, bringing a high-quality home to the inhabitants. In the whole design progress, the designer uses concise lines instead of complex decorations, giving people a deep and rough American cultural foundation. The clean white wall makes people feel as if they were bathing in the spring, fresh and comfortable with the lingering charm.

项目名称：万山庭院别墅

设 计 师：朱自权

项目地点：江苏南京

摄 影 师：逆风笑

主要材料：大理石、皮革、壁纸等

◆ 设计理念 ◆

初入门厅，挑高的中空，设计师试图给家展现一种大气浑厚的奢华生活。整个空间的色彩对比丰富，如咖啡色、米白色、以及过度的奶黄色，还有起点睛之笔的金色，丰富却不艳丽，极好地突出了整个空间的奢华感和高品质。由蓝色和大红组合而成的花纹地毯，热情奔放。搭配木质和皮革家具，恬淡舒适而不失奢华贵气，为居住者带来高质量的居家水平。设计师在整体设计过程中，均以简约的线条为主，摒弃了繁冗复杂的装饰，给人以浓厚粗犷的美式文化底蕴。净白的墙面设计又如沐浴春风，清爽舒适，经久耐看。

AMERICAN STYLE / 美式风格
NEO-CHINESE STYLE / 新中式风格
MODERN LUXURIOUS STYLE / 现代奢华风格
SOUTHEAST ASIAN STYLE / 东南亚风格

NEO-CHINESE STYLE
新中式风格

AN ELEGANT STATE
素面雅境

◇ DESIGN CONCEPT ◇

Living in the bustling metropolis, running around to do businesses and being tired of various intrigues for a such a long time, one is still pursuing for the state beyond the confusions in the world. This kind of feeling seems like life, as well as a space, which exists between the reality and void. Reality and void coexist with each other, presenting a more precious feeling. The designer transforms this kind of feeling into this case for you who is busy everyday and persist in the feeling of life yet not giving up.

项目名称：苏州旭辉美澜城

设计公司：元孚国际设计

设 计 师：胡骥

项目地点：江苏苏州

项目面积：215平方米

主要材料：大理石、胡桃木饰面、墙纸、软包、乳胶漆等

◇ 设计理念 ◇

久居繁华都市，频频奔波于各种打拼之中，疲惫于各种尔虞我诈，唯独对超然于世俗纷扰的那份追求，却不曾忘怀。这份情怀犹如人生，亦如空间，承载着真实与虚空之间。真实与虚空相生、相合，更衬情怀之可贵。设计师将这份情怀化为此项目，以献给那些日益忙碌，却不曾放弃并坚持生活情怀的你们。

Chinese style contains the fresh, elegant, restrained and modest eastern culture. The delicacy on the space layering is presented on the partition with the help of the furniture and furnishings, making the space appear a vertical feeling as well as a Zen flavor. We are attempting to use the way of reality and void to achieve a long and deep cultural flavor and make the space more elegant.

中式风格，它包含了清新淡雅、内敛含蓄、端庄的东方文化。对空间层次感的讲究，借助于家具以及陈设设计出一种虚实结合的隔断，让空间更有纵深感及一些曲径通幽的禅意。我们意图通过这种虚实之道，来达到给人一种悠远、浓厚的文化韵味，让空间更具雅致。

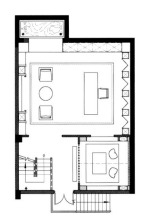
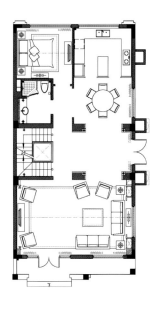

确定设计方向之后，我们在大厅地面，采用意大利树脂艺术漆塑造一幅双鱼嬉水图。这幅图由道家的太极八卦演化而来，意指浩瀚宇宙间的一切事物和现象既相互依存又相互对立的关系，如同我们所在现代社会的实和情怀中所谓的"虚"，亦如建筑空间中任何一个入人视线及行动范围中有阻隔和现实作用的空间实体，及相对应的其他元素的虚无。

在这层含义的基础之上，我们借助绘有荷花图案的儒雅屏风以及古典安逸的木隔扇门，既诠释了中国传统的朦胧美又打造空间虚实相合隔断的层次感，让光线、视线、色彩等空间元素具有延伸感。室内布局多采用对称式，在塑造空间造型上，则追求简洁硬朗的直线，来表达一种内敛、质朴的设计风格。在装饰细节上，书房意大利树脂艺术漆喷绘的山水画蕴含着含蓄而内敛的中式意境；数十个印章打造的独特天花吊顶，给人耳目一新之感。墨宝字画、匾幅、古玩、瓷器等错落有致地布置在空间中，若隐若现于隔扇之间，更添几分雅趣。

把握住中式文化的精髓，营造出雅致自然的氛围，素雅清淡的色彩和质朴自然的材质都给人带来独特的魅力。空间的主要色调为黑色、白色和原木色。黑色凸显沉静，白色尽显大方，而原木色则带来悠然之感。在这样一个隔绝外部喧嚣的静心之所，让心也更静。

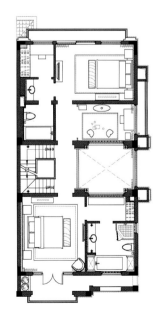

QUIET LUXURY WITH ELEGANT FLAVOR
低奢雅韵

◇ **DESIGN CONCEPT** ◇

"Simplicity is used to present the delicacy while plainness is used to build the elegance", so that the space shows its luxurious, noble and graceful beauty to the extreme. The interior is based on the soft griege tone, for the soft tone of light gray appears calm and natural among multiple colors; meticulous design techniques are adopted to present a residence with the coexistence of luxury and taste as well as life and art. Bright colors such as creamy white, rose gold and ink blue as well as the mitigation of earth color tone, let people feel the elegant and luxurious atmosphere and outline a leisure life with eastern flavor. The space is dominated by graceful, low-key and luxurious materials; on the material selection of soft decoration, the cotton and linen with good quality are matched with mercerized cloth with elements of landscape; creamy white paint and wood veneer in dark color are mainly used on the furniture, embellished with rose gold, leather and marble in some parts, making people intoxicated in the modern eastern flavor...

项目名称：掬月半山样板房

设计公司：李益中空间设计有限公司

硬装设计：李益中、范宜华、关观泉

软装设计：熊灿、王雨欣

项目地点：广东深圳

项目面积：215平方米

主要材料：蓝金沙大理石、灰金沙大理石、木地板、皮革、木饰面、墙纸、硬包、夹丝玻璃、手工地毯等

◇ **设计理念** ◇

空间以"简单呈现细腻，朴实打造优雅"，尽显奢华高贵极尽优雅之美。室内以柔和的米灰色调为主，浅灰柔和的色调，在众多色彩中淡定自然；以细致的设计手法设计一个奢华与品位共存，生活与艺术同在的起居空间。米白色、玫瑰金以及水墨蓝色等明朗的色彩再加以大地色系的沉淀，让人感受到优雅与奢华的气息，同时勾勒出一丝东方韵味的闲适生活。空间以优雅与低调奢华的材质为主，在软装材料的选择上，以有质感的棉麻搭配带有山水元素的丝光布，家具主要以米白色烤漆及深木饰面为主，局部使用玫瑰金、皮革、大理石做点缀，现代东方风情的韵味让人沉醉其中……

CLEAR ZEN FLAVOR WITH FRESH DREAMS

素禅·清梦

◇ DESIGN CONCEPT ◇

The designer in this case suggested to use traditional eastern culture with rich regional features to be the foundation and combined with western culture. The tranquil and elegant atmosphere with Zen flavor manifested the owner's steady and leisurely attitude towards life. On the selection of tones in the space, the original colors from natural materials were remained, such as brown, walnut and other dark colors, giving people a pristine and natural feel on visual sense. Based on solid wood, the designer made delicate carvings technically to build a fashionable, artful and healthy environment. Classic and fashion as well as art and elegance were combined perfectly, excavating people's demands on both body and mind of the interior environment sufficiently, while the view borrowing of the courtyard made the inhabitants be close to nature to relax themselves and enjoy the perfect home atmosphere quietly.

设计公司：良品设计公司
设 计 师：杨春雷
项目地点：河北保定
项目面积：700平方米
主要材料：石材、木皮染色、织物等

◇ 设计理念 ◇

本案设计师主张以具有浓厚地域特色的东方传统文化为根基，融入西方文化。宁静素雅的禅韵空间氛围彰显主人稳重、闲适的生活态度。空间在色调选择上保持自然材质的原色，大多为褐色、核桃色等深色系，在视觉上给人质朴、自然的本真气息；设计师以实木材质为基础，在工艺上精雕细琢，巧夺天工，塑造时尚、艺术、健康的环境空间；这里古典与时尚兼容、艺术与高雅完美结合，充分地挖掘了人对室内环境的身心需求，庭院的借景入室，让居住者亲近自然，身心放松，静享完美居家氛围。

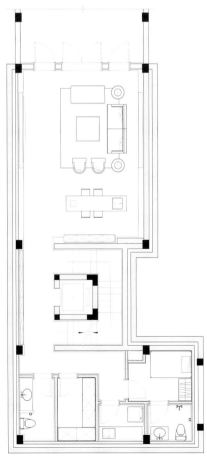
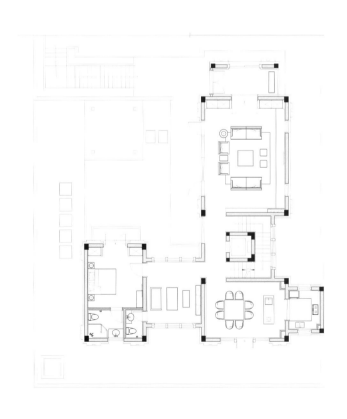

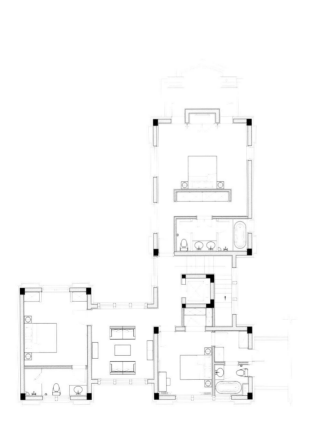
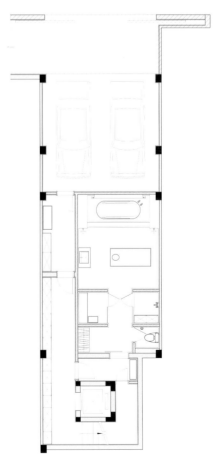

LIVING IN THE IDEAL STATE
栖居桃花源

◇ **DESIGN CONCEPT** ◇

With the rapid pace of life in modern metropolis, home seems like a shelter that could protect people's weakness. Going back home is like the returning of a tired bird or a boat. People have a psychological dependence of home as well as deeper psychological demands, such as hoping it to be poetic, tranquil or full of Zen flavor, which is the exploration of the neo-Chinese style in this case. Entering the tea room, you could feel the reclusive Zen flavor at first. With a cup of fresh tea, you can close your eyes and ruminate or overlook the beautiful sceneries outside in the afternoon. Walking slowly into the room, you will feel that large amounts of gold stainless steel lines, walnut wood and fibre are used in the interior space, making gold, khaki and yellow become the basic tone in visual sense. At the same time, beige furniture as well as decoration elements are embellished here, such as red and blue, making the overall space full of Chinese traditional colors and natural textures, which appears calm, elegant, low-key and delicate as well as some hard, neat and vivid flavor. This is exactly the balanced beauty that Pinchen Decoration is pursuing in the design.

设计公司：品辰装饰工程设计有限公司

设 计 师：庞一飞、李健

陈设设计师：张婧、夏婷婷

参与设计师：熊晓清

项目地点：重庆

项目面积：208平方米

主要材料：胡桃木、不锈钢、麻料、丝绸等

◇ **设计理念** ◇

现代都市的快节奏生活中，家仿佛是人们一处脆弱的庇护所。回家，一如倦鸟归巢，一如扁舟靠岸，人们对家具有的不仅仅是一种心理上的依赖，或许还具有更为深层次的心理需求，希求它诗意盎然，也或许宁静致远、悠悠禅意，这正是品辰对本次新中式风格设计的探索。步入茶室，禅意隐居之气扑面而来，一壶清茶，一个下午，或闭目沉思，或远眺群峦叠翠。缓缓入内，便发觉整个室内大量运用金色不锈钢线条、胡桃木质、麻料等材料，使金色、卡其色、黄色成为视觉基调。同时，利用米色家具，红色、蓝色等装饰元素穿插点缀其间，令整个空间在充满中式的古色与天然质感里，显得沉静而脱俗，低调而精致，但同样不乏些许现代硬朗、干练、生动的气息。正所谓避世而不疏世，这正是品辰所追求的设计中的平衡之美。

On the design of bedrooms, Pinchen Decoration shows a great originality. The master bedroom is themed at "lotus" while the pottery decorations have a reclusive flavor in the balanced and harmonious layout, where the ornaments such as tea sets and hand strings reflect the owner's simple and elegant interest of life. The elder's room is full of poetics in the space, while the indistinct landscape paintings on the mirror of the wardrobe convey the vacant beauty, which is more suitable for the aesthetic habit of aged people. The design in children's room is harmonious and unified with other spaces instead of echoing with the trend, while the calm color collocation and chess decoration make the children keep the quiet mood, where you could also taste the beauty far away from the hustle and bustle of metropolis. Pinchen Decoration conveys the lofty state of "while picking asters 'neath the eastern fence, my gaze upon the southern mountain rests" in this design with neo-Chinese style, where the inhabitants could understand the Tao Yuanming's philosophy of life at the same time of tasting the Zen flavor.

　　品辰在卧室的打造上更为匠心独具，主卧以"莲"为主题，陶艺装饰在平衡而和谐的布局里别有一派隐居之气，茶具、手串等摆件无一不体现主人的简美生活情趣。老人房则充满空间诗学，衣柜镜面的山水意境画若隐若现，传达出空灵之美，更符合长者的审美习性。儿童房的设计并没有一味附和潮流，而是同其他空间和谐统一，安静的色彩搭配、棋艺装饰，让孩子保持安静的情绪，同样能够品味幽玄之美，远离都市的喧嚣。品辰欲以本套新中式设计，传达"采菊东篱下，悠然见南山"的高远意境，住户在品味隐居禅意的同时，便懂得陶渊明这位朴素"大富翁"的生活哲学。

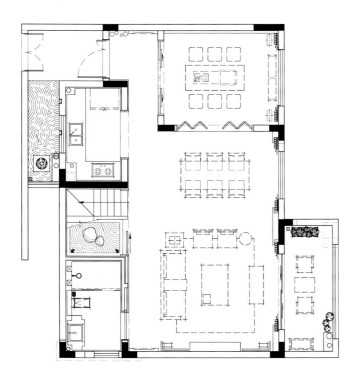

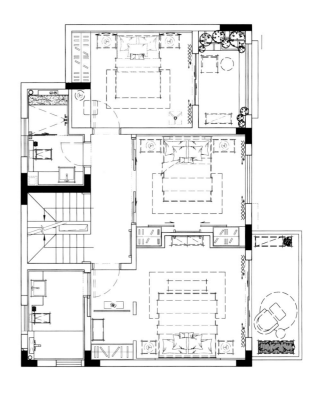

LITERARY FRAGRANCE IN THE MAGNIFICENT MANSION
书香致远 屋华天然

◇ DESIGN CONCEPT ◇

On the premise of knowing the intentions of the owner, the designer blended traditional cultural understanding into the modern design. Through the recreation of traditional culture, the designer combined the books, calligraphy, plum blossom, asparagus fern and other classical art elements that rooted in Chinese traditional culture with modern design languages, creating an elegant and distant atmosphere.

设计公司：福州宽北装饰设计有限公司
设 计 师：郑杨辉
项目地点：福建福州
项目面积：272平方米
撰　　文：vivan

◇ 设计理念 ◇

本案设计师在洞悉业主意愿的前提下，将传统文化理解吸收到现代设计当中去。通过对传统文化的再创造，把根植于中国传统文化的书籍、书法、梅花、文竹等古典艺术元素和现代设计语言完美结合，营造一种高雅悠远的氛围。

At the same time, diversified techniques were used to interpret traditional cultural elements again to make it merged with the new environment and new modeling, which used the space interface as the carrier, creating a space full of sense of culture, beauty and interest. The cultural and decorative features of these elements brought higher levels of artistic conception into the space, making the interior decoration have modern sense as well as exude historical and traditional atmosphere, which was sentimental yet full of connotation, satisfying the owner's beautiful pursuit and wish of the hereditary family tradition "poems and manners".

与此同时,设计师还通过多元化的手法对传统文化元素进行新的演绎,让其与新环境新造型有机融合在一起,以空间界面为载体,创造出富有文化、美感和情趣的空间。这些元素本身具有的文化性、装饰性,也给空间场所带来了更高层次的意境,使室内装饰既具有时代感,又能散发出历史传统气息,既富有情调,又不失意蕴和内涵,充分满足主人期望"诗礼传家"的家风代代相传的美好追求和愿望。

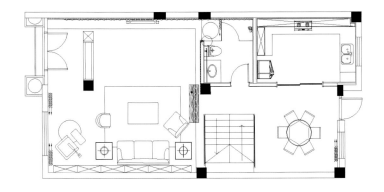
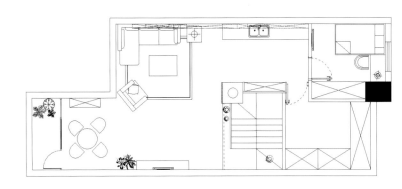

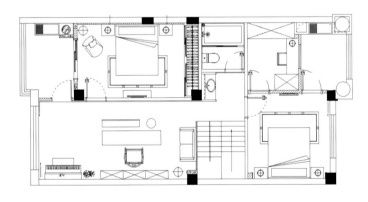

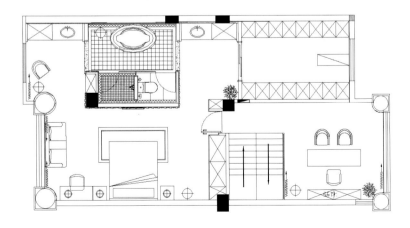

人们常说："知书达礼"。人的气质需要书的滋养，同样的道理，家的装修不在于"看得见"的奢华，而在于能否锻造出空间的内涵和气韵，正所谓"最是书香能致远，'屋'有诗书气自华。"当人、书、空间三者之间建立起一种紧密的联系，空间就不再是一个纯粹的物质存在，"书"也超脱了"装饰物"的范畴，变成了空间的灵魂和支点。就像本案的业主，他是小学老师，非常喜欢书。对他而言，书是最佳的品味代言。所以他特别强调设计师要帮他打造一个富有"书香气"的家居空间，让这个家的美不再限于表面，而是符合主人对精神文化的更深层次的追求。

家是人们赖以生存和活动的最重要的空间环境，无论是高楼别墅，还是小小平房，或多或少都会带着主人的气质和品味。本案设计师在充分了解业主需求的基础上，精心调配出妥适的格局。净白空间里，由"书"元素延伸出的各种造型、手法，营造出灵气盎然的人文意境。

客厅的设计充分应用了"书"元素这一有意味、有内涵的形式。客厅的地板通过光面与哑光面瓷砖的结合来形成一种独特的视觉效果，设计师特意将其切割成不同大小的"书脊"形状，跟墙面形成一体式的造型，而且与"书盒"外观的厨卫连体空间形成呼应，给人浑然一体的构图美感。书架也是采用异形拼贴手法，富有动态感。餐厅的吊顶被设计师有意"拔高"，使空间更通透，古朴谐趣的壁画默默倾诉着"家和"、"有余"的中华情结。餐厅旁边的玻璃推拉门既可以隔离油烟，又放大了空间的视野。另外，设计师还从传统水墨画艺术中汲取灵感，对客餐厅空间进行虚实结合、张弛有度且富有层次感的分隔，通过其独特的艺术形象和文化性，将更多的信息附加于空间界面之上。

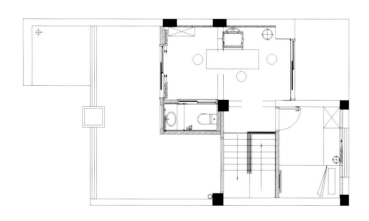

灯光的设计也是本案的一大特色。客厅、餐厅、卧室、会客室等大部分功能区域的吊顶所用的灯具全部采用隐形灯与射灯相结合的形式，使空间更显简洁、素净、内敛。卧室采用不同色阶的黑白灰，调和出一个极简的时尚空间，大面积的实木铺陈，给人舒适温馨的审美体验。

地下室的设计简明通透，并通过天窗的运用，引入花园的自然光，同时搭配玻璃、陶艺、麻布、草编等材质，营造出一种纯净如水的空间意象。

空间适度留白能成就美的篇章，而在适合处填空，也能为空间提升价值。本案设计师在适当的角落利用陶瓷艺术品、绿植给空间"填空"，充分发挥这些景观小品的形态美，打造空灵雅致的环境效果，让人置身室内也能享受户外庭院的美景和悠闲的氛围。

TIME IN THE CORNER OF THE CITY
城隅时光线

◆ DESIGN CONCEPT ◆

Wood is the design theme in this case while elements with Zen flavor are flowing throughout the whole space, where movement and quietude are just separated. You can enjoy a leisurely afternoon in a totally carefree mood. We live in an apathetic emotion and spiritless thought, thus we need natural elements to embellish and wake up our boring life. Designers use a plain design to let us see the ordinary and sincere emotions in the mincing world. The blossom of flowers is endless, while the memories will be disappeared with times. Say goodbye to the corner of the city, then the only thing you feel is nothing at all.

设计公司：品辰装饰工程设计有限公司
设 计 师：庞一飞、颜飞
参与设计师：叶乙霄
项目地点：重庆
项目面积：320平方米
主要材料：深色系木做、木纹石、手绘墙纸、木地板等

◆ 设计理念 ◆

以木为设计主题，禅意的元素肆意整个空间，静与动只一线之隔。在一种全然悠闲的情绪中去消遣一个闲暇无事的午后时光。我们生活在已然麻木的情感与死气沉沉的思想中，需要自然元素的点缀来叫醒无趣的生活。设计师用朴质的设计让我们在矫饰的世界中看到平凡真挚的情感，荆棘厮杀不枉轻狂年少，高不胜寒不若一室暖光等候。看不尽繁花似锦，时光荏苒往事终归墟。城隅再见，却只觉云淡风轻。

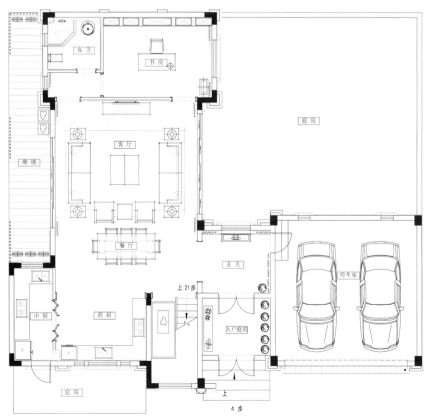

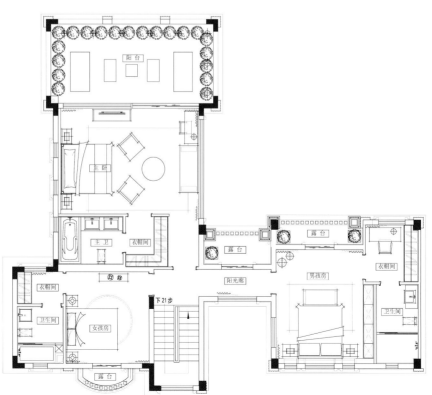

CHINESE RESIDENCE WITH ZEN FLAVOR
中式禅居

◆ DESIGN CONCEPT ◆

The atmosphere of the overall space is elegant and pristine with Zen flavor. Although they belong to the same building, they are three staggered-floor suites in fact. Beige, sage green and water blue make the space interpret and extend different gradation, while the materials and soft decorations with different textures make these suites harmonious with rich layers. You will not feel uninteresting or boring even though entering so many rooms. The sunlight shines through the French window, making people feel warm throughout the body. If you make a pot of tea and sit to enjoy the leisure far away from the hustle and bustle, it will surly be the most pleasant thing. People with different ages, regions and preferences could feel that there could be a home.

项目名称：无锡灵山小镇•拈花湾系列作品之样板区

设计公司：禾易HYEE DESIGN（原HKGGROUP）

设 计 师：陆嵘

参与设计：李怡、卜兆玲、王玉洁

项目地点：江苏无锡

项目面积：566平方米

◆ 设计理念 ◆

整体空间氛围素雅清丽、古朴禅意、丝竹交错。虽同处一栋，却是相互交织的三套错层套房，米黄、灰绿、水蓝三个不同的色调，空间演绎又延伸出不同灰度、不同肌理的材质和软装，使得这三户语调统一、层次丰富。即使不断进入这么多个房间，也不会觉得无趣或乏味。阳光由落地窗撒入，整个人都觉得暖洋洋的。这时候沏上一壶茶，静坐着感受这远离尘嚣的安逸和优闲，怕是人生最惬意的享受了。不同年龄、不同地域、不同喜好的人对此都能感同身受的知道在这个地方，能筑一个家。

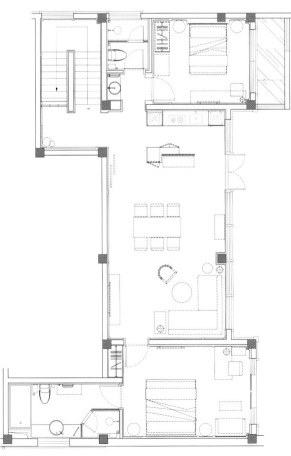
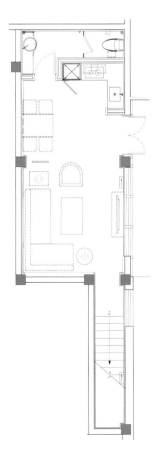

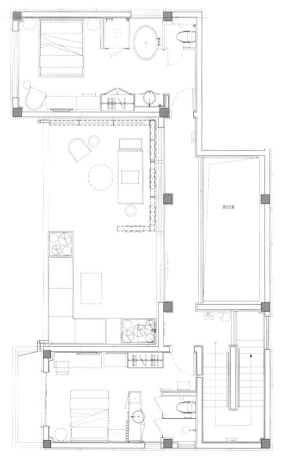
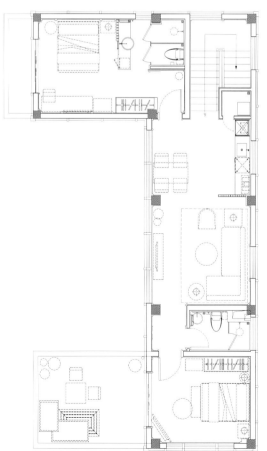

THE FRESH ZEN
清雅禅音

◇ DESIGN CONCEPT ◇

Thatched roof, log furniture, plain colored fabrics are all careless and natural without any decoration. The warm sunlight shines through the solid wood partition with a chiffon, which seems like an elegant and calm fairy land where a heart could be infiltrated to be crystal clear and smooth like a mirror without any dust. Floristic decorations are more vivid and vital than any other decorations, while the birds on the branch and the dried plants in the painting are collocated to enrich the space as well as bring a new visual impression. It is elegant with Zen flavor that could reach people's inner heart. The broad layout, elegant and concise arrangement and the leisurely temperament throughout the space could make visitors intoxicated. It seems that the essence of life should be like this, which is quiet and comfortable.

项目名称：无锡灵山小镇•拈花湾系列作品之样板区

设计公司：禾易HYEE DESIGN（原HKGGROUP）

设 计 师：陆嵘

参与设计：李怡、卜兆玲、王玉洁

项目地点：江苏无锡

项目面积：303平方米

◇ 设计理念 ◇

　　茅草屋顶、原木家具、素色面料，一切都是那么的不经意，天然去雕琢。暖暖的阳光透过夹着薄纱的实木隔断，仿佛置身在了一个清丽如水，沉定如钟的桃源幽世，一颗心也被浸润得晶莹剔透，平滑如镜，不惹半点尘埃。植物装饰比任何饰品更具有生机和活力，画中的树枝小鸟与干枝的植物搭配在此，或虚或实，或静或动，不但丰富了空间，还给人们带来全新的视觉感受，空灵脱俗，直达心灵的禅音。开阔的格局，清雅简约的布置，空间里透出的那份悠然气质，是观者可以欣然入境的。仿佛生活的真谛就该如此，淡薄、安适。

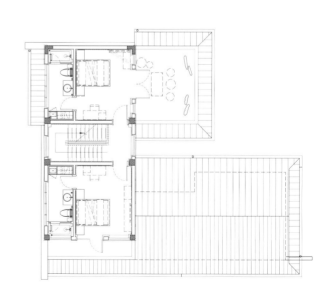

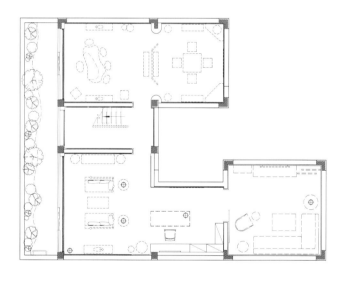

THE FLOWING FRAGRANCE
暗香浮动

◇ DESIGN CONCEPT ◇

A woody perfume called "TERRE" of HERMES is still highly praised by modern people. Wooden furniture exude a dry and warm atmosphere in the air with implicit fragrance, making the generations reminiscent of them. On account of the owner's love on solid wood, African rosewood is largely used in this case to produce overlength and superwide solid wood floor, gratings, wardrobe, wash basin and most removable furniture. In order to avoid the excessive heaviness of the atmosphere due to large amount of wood veneer, the wall surface is dominated by white or gray wallpaper, while at the same time, sofa, curtains and other accessories are based on griege tone to mitigate the visual impact brought by large area of wood color. The gentle wood color appears more calm and textured in the white interior space.

设计公司：卓新谛室内空间营造社

设 计 师：卓新谛、卓友彬

项目地点：福建福州

项目面积：280平方米

主要材料：非洲花梨、复古砖、墙纸、硅藻泥、艺术玻璃等

◇ 设计理念 ◇

HERMES 一款木质调主打的"大地"香氛，至今为人推崇。暗香浮动，空气中，木质家具散发出干燥的、温暖的气息，总让一代又一代人怀念。由于业主对实木的热爱，本项目大量选用非洲花梨木，用来制作超长超宽的实木地板、格栅、衣柜、厨柜、洗手台及大部分移动家具。为了防止大量的木面，让空间氛围显得过于沉重，墙面设计以白色调或者素色的灰调墙纸为主，同时沙发、窗帘等饰品也以米灰色调做主打，缓和大面积木色带来的视觉冲击。温润的木色在白色的室内显得更加沉静，有质感。

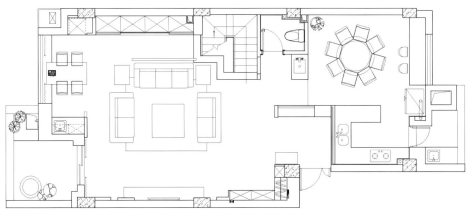
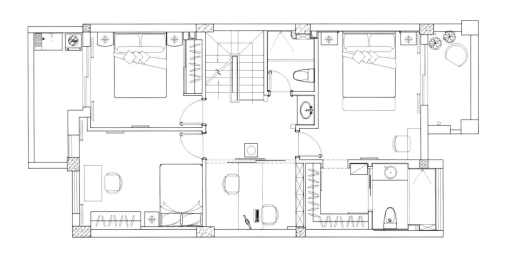

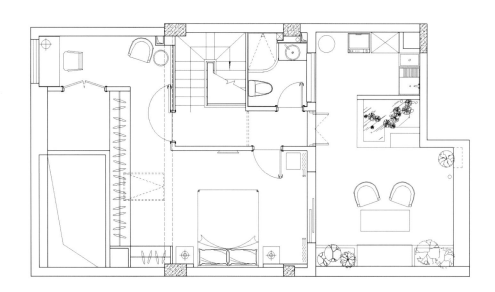

SCULPTING IN TIME
雕刻时光

❖ DESIGN CONCEPT ❖

When life becomes the most luxurious time, you could enjoy the leisure with a cup of English afternoon tea quietly. There is a kind of leisure with luxury as well as simplicity with taste and casualness, which is full of exotic flavor and rich exotic cultural atmosphere, conveying the diversity of life. The urban new aristocrat who are living in the armored concrete with the pressure of life and struggle for existence, receive the "warm embrace" on both body and mind, which becomes the rest place of heart.

项目名称：长春·中海寰宇天下示范单位
设计公司：PINKI品伊创意集团&美国IARI刘卫军设计师事务所
设 计 师：刘卫军
参与设计：梁义、张罗贵
陈设设计：PINKI品伊国际创意集团&知本家陈设艺术机构
陈 设 师：李莎莉
项目地点：辽宁大连
项目面积：85平方米
主要材料：实木地板、墙布、大理石等

❖ 设计理念 ❖

当生活成为最奢侈的时光，你可以静静地感受一份禅茶带来的惬意。一种休闲中带有奢华、简约中带有品味、随意，处处充满着异域风情的意境和浓郁的异域文化气息，传达着生活的多姿多彩。使聚居于钢筋水泥为支持的现代都市新贵在接受生活的压力和生存的竞争中，能在这里让身心得到交流的"温暖怀抱"，使之成为心灵的小憩地。

FEEL QUIETLY
静悟

◇ DESIGN CONCEPT ◇

We always hope to have a home where we could calm down and feel life slowly, which is a real harbor of mind. Thus, this case is created. Quiet is a kind of feeling, when we stay at the elegant house with a cup of tea, the thoughts are flying; quiet is a kind of beauty as well as a realm, which is an insight after understanding life; quiet is a kind of feeling, which returns into the tranquil beauty at last. If we could remain a quiet space in life, we could keep our inner hearts as well as step away from the fretted living conditions at the moment, for we could feel the essence of life from the world full of hustle and bustle.

设计公司：CEX鸿文空间设计有限公司
设 计 师：郑展鸿、刘小文
项目地点：福建
项目面积：300平方米
主要材料：银貂大理石、壁画、木饰板、原木、麻布壁纸、麻纱窗帘等

◇ 设计理念 ◇

我们总是希望，我们能有一个可以让自己静下来慢慢感悟生命的家，一个真正的心灵港湾。所以，就有了这个作品。静是一种感受，当我们独处在静雅陋室的时候，品着一杯香茗，点上一株檀香，任思绪随着袅袅轻烟飘飞；静是一种美，也是一种境界，是看穿人生沉浮的一种顿悟；静是一种悟，怒云狂风，终为雨露，归于静美。倘若我们能够在自己的人生里存留下宁静的一个空间，不仅能使自己的内心得以寄留，也能使我们从焦燥的生活状态中走出，在喧闹忙碌的世界里分出一点闲暇去体悟内心的生命本源。

The layout is open and free in this case with clear and ordered generatrix, the expression of the scenery with steps is used to bring people inside. After communicating with the owner, we use the simple style from Tang and Song dynasty instead of the common furniture in Ming and Qing dynasty. On the details, we try to be more clean and neat, simplify excessive renderings, interpreting all of these with clean surfaces and lines. The living room, dinning room, chess room and staircase at the first floor are all opened, partitioned simply with door pockets and lines, making all the spaces apparent. No excessive gorgeous elements are used on the materials, instead, a number of pristine materials such as rubble, raw wood and flagstone are adopted, opening the layering of the space through the lamp lights and exclusive natural lights. On the accessory, small green plants, books, gold bowls, old vats, plum blossom and small chandelier as well as the painting of *Bamboo and Stone* from Banqiao Zheng are used to present the tasteful flavor of the space.

本案在空间布局上大开大合，整个动线明朗有序，采用一步一景，步移景生的表达手法，引人入胜，渐入佳境。在和业主做深入交流后，避开常规中式的明清花格、明清家具，更多的采用唐宋盛行的简约风，空间在细节上的处理尽可能的干净利索，简化掉过多的渲染，而用干净的面与线来诠释。本案把一楼客厅、餐厅、棋牌室、楼梯间全部打开，用了门套和线条的形式做了简单的分割隔断，让所有空间的动线若隐若现。本案在材质上没有用过多华丽的元素，而是更多的用一些质朴的材质，毛石、原木、青石板，借上层次分明的灯光和本户型拥有的天然采光，把空间的层次完全拉开。在配饰上，用小绿植、书、金钵、古缸、梅花、小吊灯及郑板桥的《竹石图》把空间的韵味展现的耐人寻味……

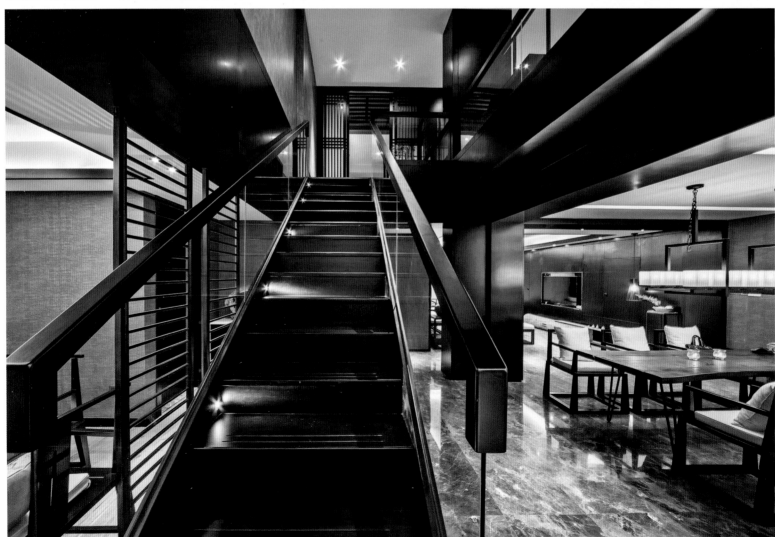

AMERICAN STYLE / 美式风格
NEO-CHINESE STYLE / 新中式风格
MODERN LUXURIOUS STYLE / 现代奢华风格
SOUTHEAST ASIAN STYLE / 东南亚风格

MODERN LUXURIOUS STYLE
现代奢华风格

THE PURITY
繁华落尽见真淳

◆ DESIGN CONCEPT ◆

From the view of furnishing, designers start a design rhythm in a modern urban living space with their sharp handling on colors and concise technique of expression, interpreting the unique charm of the interior space. At the same time, fashionable and modern elements are blended in life to emphasize the quality and sense of art.

This space is dominated by off-white and gray, while lake blue and coffee are used as the foreground colors to interpret the fashionable and elegant space, whilst the lemon yellow floriculture adds the vitality. Natural lusterless materials and high-gloss materials are combined in the living room to make the space full of elegant home flavor yet pervade modern taste. The black mosaic background wall jointed with dazzling seashells seems like the bright stars, enhancing the luxurious sense of the overall space. Typical modern elements such as the ebony dinning table, clear lines and light colors, create a dinning room which echoes with the living room.

项目名称：中航城翡翠湾143户型
主案设计师：郑树芬（Simon Chong）
参与设计师：杜恒(Amy Du)、陈洁纯（Holiday Chan）
项目地点：湖南岳阳
项目面积：139平方米

◆ 设计理念 ◆

设计师以其对色彩的敏锐把握，从陈设的角度，开启了现代都市生活空间的设计节奏，以简单的表现手法写意出室内空间的独特魅力。同时，将时尚现代元素融入生活之中，强调品质与艺术感。

空间以米白灰色为主、运用了湖蓝色、咖色作为前景色来演绎出时尚、雅致的空间，用柠檬黄的花艺点出空间的活力。客厅的设计将天然哑光材质与高光亮泽材质进行有机结合，使空间充满雅致的居家风情，又渗透出现代时尚的韵味。由晶亮贝壳拼接而成的黑色马赛克背景墙，宛如夜幕中璀璨的繁星，光彩夺目，提升了整个空间的奢华感。黑檀亮光木饰餐桌，清晰的线条，淡雅的色彩，这些经典的现代元素塑造出一个与客厅相呼应的餐厅。

The bright abstract painting at the end of the aisle connect all the spaces, the rich expressive force as well as the unified and harmonious tone of which gives people an enjoyment of high-quality home life. Each corner tells that life is deserved to be wonderful. The gorgeous dark red endows the master bedroom an elegant and luxurious temperament, where the coffee leather and the painting with black and white simple lines are used as the foreground colors, collocated with metal accessories and orange floriculture, making the whole space exude a harmony of colors and quality, which is rich and ordered.

在过道尽头的亮黄色抽象画将各空间承上启下，丰富的表现力和统一和谐的色调，给人高品质居家的享受，每一个角落，都在诉说生活就应如此美好。瑰丽的深红色赋予主卧优雅的奢华气质，以咖色的皮艺，黑白简约的线条的挂画为前景色，搭配亮面金属质感饰品和橙色花艺为辅助色，使整个空间流露出色彩与品质的和谐，丰富而有序。

THE SUNBATHE
沐阳

◈ DESIGN CONCEPT ◈

You couldn't imagine how it looks like before you enter this fantasy castle. The designer builds space expressions which could not be duplicated for the owner exclusively, making people feel the dizzying visual impact as soon as they enter the room. There's no deliberate style here yet funny and elegant with leisure. The special texture on the top of the aisle builds the rough and concavoconvex sense like the grotto and cliff, added with puerto design in the same technique, reflecting the attitude towards life of back to nature that the designer and owner are pursuing. The octagonal ceiling in the dinning room makes the special corner individual without any abrupt impression. The corners under the staircase of the second floor are decorated delicately where the sunlight could shine through the window warmly onto this fairyland. The bathroom with black and white tone is elegant and clean, while the freehand tree on the wall makes the space interesting. When the sun shines aslant and the shadow drops on the tree, it makes people feel glad as if there's a flower blossom in their heart no matter the light is strong or not.

设计公司：云想衣裳室内设计工作室
设 计 师：连君曼
www.tofree.com.cn
项目地点：福建福州
项目面积：284平方米
摄 影 师：周跃东
主要材料：水曲柳面板、杉木、松木、乳胶漆、仿古砖、布艺、文化石等

◈ 设计理念 ◈

在没有进入这个奇幻的城堡之前，你一定想象不到它的模样，设计师为屋主专属打造了不可复制的空间表情，入室一瞬间有着目不暇接的视觉冲击感。这里没有刻意的风格定位，却自成一派，风趣优雅，闲情自在。过道顶部的特殊肌理打造岩洞石壁般的粗糙凹凸感，加上同样手法的垭口设计，反射出设计师与屋主追求回归自然本真的生活态度。餐厅的八角吊顶让这个有着特殊转角的区域没有突兀感且个性十足。二楼楼梯转角下面的畸零角落也被装点的别有洞天，阳光凭窗而入，暖暖地照着这乐园。黑白格调的卫浴间唯美、纯净，墙壁上的手绘树让这个空间妙趣横生，阳光斜照入室，光影落在这树上，强烈的或柔和的，都让人美的心中开出一朵花来。

MIRROR·CLEANNESS
镜·净

◇ DESIGN CONCEPT ◇

This case has a strong sensitivity on fashion, pursuing for the group design that modern comfortable life needed. Based on modern Hongkong style, the designer broke traditional design techniques and abandoned inflexible design forms to connect the living room, kitchen and bedrooms, which enriched the visual sense to the maximum, adding the liquidity of the space. The overall decoration was based on simplicity and cleanness, especially the handling on details. Considering the demands of the client, the designer customized a lot of metal parts, reflective glass and other furniture decorations, giving a simple yet fashionable feeling. Dominated by a warm tone with simple lines, the space was concise and full of taste. Metal colors and lines were used to create the splendid and magnificent luxury yet not too extravagant, instead, a sense of quality was embodied, creating a warm and leisure atmosphere.

设计公司：VBD华优室内

设 计 师：翁瑞栋

项目地点：浙江绍兴

项目面积：435平方米

摄 影 师：林德健

主要材料：墙纸、木饰面、大理石、反光玻璃等

◇ 设计理念 ◇

本案针对时尚有很强的敏感度，针对追求舒适现代生活居住需求的群体而设计。以现代港式为主，颠覆传统的设计手法，抛开死板的设计形式，将客厅、厨房、卧室贯通起来，最大程度地丰富了视觉观感度，也增加了空间的流通性。整体装饰以简洁干净为主，细节上的处理尤其到位，考虑到客户的需求，定制了许多金属件、反光玻璃等家具装饰，给人一种简约而不失时尚的感觉。该项目以暖色调为主，线条简单，空间简洁而富有品位。以金属色和线条感营造金碧辉煌的豪华感，不会太过于奢华，反而体现出一种品质感，营造出一种温暖、休闲的氛围。

From the point of design, this case was adopted with symmetrical layout on the basis of keeping the reasonable generatrix on the plane to show the theme of conciseness, brightness and comfort. Stainless steel, wood veneer, marble and other materials with reflectance were used on the façade to embody the fashionable and dynamic quality of life. Wallpaper, hark decoration and soft decoration were applied to build a living environment full of comfort, making people have the recognition on this modern and concise design with artistic texture.

从设计角度来看，该项目在平面上保持动线合理的基础上运用对称的布局方式，表现简洁明快舒适的主题。立面上主要为不锈钢、木饰面、大理石等具有一定反光度的材料，表现时尚和动感的生活品质；运用墙纸，硬包和软包，打造极富舒适感的生活意境，让人们对现代简约、艺术质感的设计产生认同感。

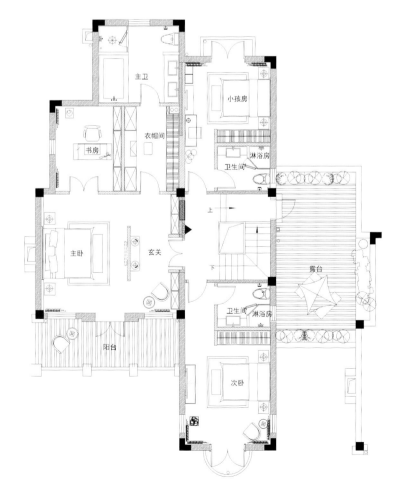

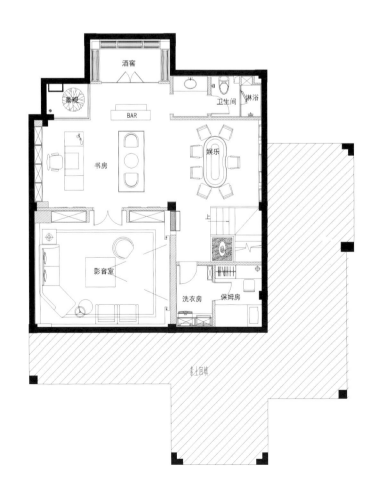

THE FUTURE HOME
未来之家

◇ DESIGN CONCEPT ◇

The target consumers are young people who pursue for fashion and trend as well as emphasize on the practicability of the living space. On the basis of modern simple style, the designer mixes other styles to add new elements into the space with full interests. This house type simulates a home with four family members, attempting to build a modern, individual and fashionable spatial atmosphere on the basis of modern style, which is centered on the interests, habits and lifestyles of the inhabitants.

项目名称：福建龙旺理想天街 A1户型
设计公司：陈铌设计
设 计 师：陈铌
项目地点：福建福州
项目面积：82平方米
摄 影 师：施凯
主要材料：大理石、白色半亚木饰面、黑胶网纹玻璃、黑钛不锈钢、镜面玻璃、硬包、墙布等

◇ 设计理念 ◇

SOHO的消费人群定位大多是以年轻人为主，追求时尚与潮流，非常注重居室空间实用性。设计手法上在现代简约基础上又辅以混搭，给空间注入新的元素，富有趣味性。此户型模拟一个四口之家，在现代风格的基础上，围绕居住者的兴趣爱好及生活方式，打造一个现代、个性与时尚的空间氛围。

Entering the living room, the furniture you will see come from the typical type of an Italian top brand MINOTTI IT. They are modern, concise and practical with individual fashion. The relaxed and comfortable blue sofa jumps into our eyes, while the embellishment of orange enhances the lively atmosphere in the space, catering with modern young family's demands on living space. The dinning room continues the overall tone and modeling in the living room, creating a relaxed dinning atmosphere.

走入客厅部分，所看到的家具是选用意大利顶级品牌MINOTTI IT的经典款式。现代简约实用的家具形体，个性时尚，轻松舒适的蓝色休闲沙发进入我们眼球，橙色的适当点缀，提升了空间的活泼氛围，迎合了现代年轻家庭对居住空间的需求。餐厅延续客厅的整体色调及造型，营造一个轻松的就餐氛围。

The master bedroom is a place simulated for a young couple with higher education who pursue for a romantic, artistic and funny space, The furnishings and posters embody their careers and interests. The children's room is designed for a pair of 8-year-old boy twins, themed as Captain America, a hot movie in recent two years, while the bunk bed makes it have more activity space for the children.

主卧，模拟的是受过高等教育的年轻夫妻，追求空间浪漫、艺术及情趣。空间的摆设及海报，体现了他们的职业及生活兴趣。儿童房是一对8岁的双胞胎兄弟，以近两年非常火热电影的《美国队长》为主题，上下床铺使房间多了一些小孩的活动空间。

THE ELEGANT LIFE
典雅生活

◇ DESIGN CONCEPT ◇

The using of symbols of element in the space design adds a little atmosphere of eastern aesthetics, collocated with meticulous artworks, soft and enchanting lighting effect and the fabric of silk, cotton and linen, highlighting designers' unremitting pursuit of art-deco style and the elegant living environment.

项目名称：路劲集团（常州）X1户型样板房

设计公司：矩阵纵横

主创设计：王冠、刘建辉、于鹏杰

参与设计：吴比、周晓云

项目面积：285平方米

主要材料：灰木纹大理石、灰橡木、印花皮、木地板、壁纸、地胶等

◇ 设计理念 ◇

空间设计中元素符号的运用为整个空间增添了一丝东方美学的气息，配以一丝不苟的艺术品、柔和妩媚的灯光效果和丝绸、棉麻的针织布艺，充分突显了设计师对装饰艺术风格及典雅居住环境的不懈追求。

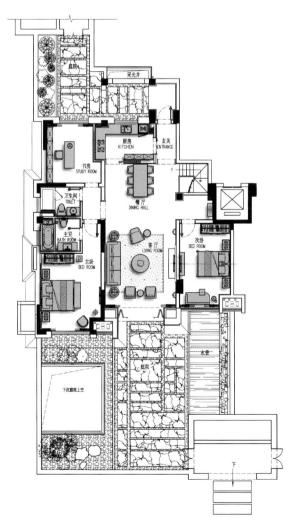

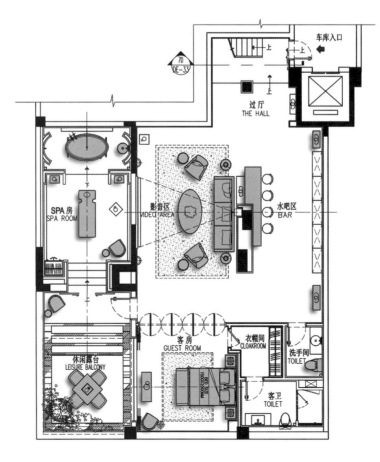

A RHAPSODY IN ILLUSION
幻景狂想曲

◇ DESIGN CONCEPT ◇

On the basis of following the beauty of traditional classical symmetry and proportion, designers blended in modern views that simplified traditional lines and complex modelings to convey an unconventional elegant temperament by using large amount of soft and elegant colors and materials, whilst the soft decorations with good texture were added to embellish the fashionable highlights. Furniture in ARTDECO style was combined with Chinese flower arrangement, the simple structure, elegant temperament and the classical oriental beauty of which were matched perfectly, presenting a unique and harmonious design aesthetics. The sunshine-like orange color flowed through the graceful space in gray tone, while the sense of purity created a leisure and comfortable living atmosphere. Wandering here, you could feel the fleeting time slowly, and taste the quiet years carefully.

陈设设计公司：北京中合深美装饰工程设计有限公司

设 计 师：郭小雨、石哲

项目地点：北京

项目面积：850平方米

主要材料：壁纸、地毯、装饰画等

◇ 设计理念 ◇

设计师在遵循传统古典对称和比例美感上的基础上，融入以简制繁的当代观点，简化传统线条和造型的繁复，大量使用柔和素雅的色彩和材质，传递一份不落俗套的优雅气质，并悉心加入质感软装衬托出时尚亮点。ARTDECO风格的家具与中式插花相结合，简单的构造、典雅的气质与东方的古典之美不谋而和，呈现出独特和谐的设计美学。阳光般的橙色穿行于优雅的灰调空间中，一抹纯粹，营造出惬意舒适的生活气氛。倘徉于此，淡看流年烟火，细品岁月静好。

FRAGRANCE THROUGH TIMES

时光倾吐芳华

◆ DESIGN CONCEPT ◆

The layout and facilities in this case are similar to those in hotels, collocated with decorative lightings in various modelings, creating a modern and luxurious residence.

The noble temperament appears as soon as you enter the living room, while the chandelier made of different golden rose annuluses is generous and steady, collocated with the tea table in tree root modeling with antique gold foil and acrylic, rendering the heavy sense of the environment. Modern design materials and soft materials are combined on the wall, matched with the floor lamp and the halogen lamp on the ceiling, making the space successive, which embellishes the mobility of the space and creates the warm feeling of the family reception room.

项目名称：广东•东莞龙泉豪苑2#2101样板房

设计公司：柏舍励创•柏舍设计

项目地点：广东东莞

项目面积：500平方米

主要材料：大理石、玫瑰金镜钢、夹丝玻璃、水晶灯等

◆ 设计理念 ◆

本案的布局和设施与酒店相仿，搭配各种造型的灯饰，营造一个现代雍容的居住空间。

一进门的客厅便尽显贵族气质，由多层不同半径玫瑰金圆环重叠的吊顶水晶灯，大气沉稳，配以树根造型茶几，以仿古金箔与亚克力做旧，渲染出空间环境的厚重感。墙面运用现代设计材料与软面材料相结合，配合落地灯光与天花板卤素灯的布局，使空间有分有合，层层递进。衬托出空间的流动性，营造出家庭会客厅的温馨感受。

Rose gold is also adopted on the decorations of the water bar, while rectangle becomes the theme here, where tough and simple lines echo with the luxurious feeling of the overall space such as on the showcase, the ceiling, the lamps and the water bar. The design in the bedrooms is steady and generous, for example, Chinese pane element is extracted on the main wall, while the lighting design surrounded enhances the artistry of the space. Through the details, you can feel the warm and gorgeous flavor of the whole space, such as rose gold steel mirrors, the delicate chandelier decoration of the international chess and several decorative paintings of streetscape embellished among them.

水吧区的装饰同样采用玫瑰金，长方形成为水吧区设计的主题，橱窗、吊顶、台灯、水吧台，用硬朗简单的线条呼应整体空间的雍华感。卧房设计稳重大气，主幅墙提取中式窗格元素，一方一圆，一天一地的灯光设计环绕整个空间，提升了空间的艺术性。玫瑰金镜钢的多处装饰，书房国际象棋的水晶挂饰精致而充满味道，点缀其间的几个街景装饰画，细节之处感受整个空间温暖华丽的神韵。

A LIFESTYLE DESTINATION
寻找生活的归宿

◇ DESIGN CONCEPT ◇

A fascinating life must be brimming with art, music, intrigue and romance...

The value of design is changing the quality of life, creating a new lifestyle and finding the resonance of spirit and life.

This house type is defined as neo-oriental style with a steady and heavy tone, while the furniture collocation is mainly based on the Italian brand B&B, which is noble yet not showy, simple yet not plain, appearing the features of the overall space. Time is the principal axis while classic is the essence that being slowly brought here. Shanghai flavor is the narration of culture. On the space layout, not too many elements of old Shanghai are adopted, while the extraction of the design and the abstract oriental flavor are expressed in the space, where you could see the spirit with quality.

项目名称：宝华紫薇花园E户型

设 计 师：连自成

参与设计：金李江、江燕

项目地点：上海

项目面积：160平方米

主要材料：胡桃木、毛面砖、铜艺栏杆、大理石、陶瓷、铜镀铬等

◇ 设计理念 ◇

所谓迷人的生活必定是充满着艺术，音乐，刺激和浪漫……

设计的价值就是改变生活品质，创建新的生活方式，觅得精神与生活的共鸣。

此户型是新东方的风格，整体基调稳重浑厚，家具的搭配以意大利品牌B&B为主，高贵而不炫耀，简单但不平淡，是整体空间呈现的特征。以时间为主轴，经典是缓缓带入的精髓。上海风情，是一个文化的叙述，在空间的布局上，并没有过多采用老上海的元素，设计的提炼，抽象的东方气韵被表达在空间中，所能见到的是品质追求的精神。

People's pursuit on quality is the standard of an era. The choice of colors is unified, such as the classic with Chinese flavor and the metal decorations with modern sense, which are not antipathetic. A few artworks of color decorative painting in modernism are the highlights which not only enhance the overall temperament, but also neutralize the heavy sense of the oriental colors. Designers combine the past and the present in the same space-time, where the complex is connected with the simplicity, such as the petty elegant Chinese elements, the texture and color with the features of the marble era and the modern home furnishings, attempting to interpret the owner's taste and wisdom.

What we are looking for in life is the destination of ourselves, so that we use designs to bring people into the real living conditions.

　　这是一个时代的标准，人们对品质的追求。色彩的选择有统一调性，中式风情的古典，现代感金属材质的装饰品，于空间中相间搭配却并不突兀。以一两件现代主义色彩装饰画的艺术品画龙点睛，不仅提升整体气质，也是缓和东方色调的厚重。将过去和现代在同一时空交融，一些细碎雅致的中式元素，大理石时代特征的纹理色泽，现代的家居装饰，繁复与简洁对接，意图去阐述居者的品位和智慧。

　　在生活中寻求的是自身的归属，用设计将人们带向真正的生活状态。

TIME COLLECTION
时光收藏

◇ DESIGN CONCEPT ◇

SOHO represents a working mode which is more free, open and flexible. The coming of the network era makes SOHO become a living mode that young people are pursuing for, which is a vogue combination of living and working. In this flexible space full of imagination, the texture of the materials and the delicate expression of design techniques are emphasized to endow the fashionable and elegant temperament. It shows the owner's preference on clocks in the embodiment of furnishing art, which is the collection of time! The basic tone of the space is restrained and low-key like the mellowness of wine, bringing a wonderful, excessive and happy life with an international metropolis flavor!

项目名称：福州奥体SOHO样板间

设计公司：深圳张起铭室内设计有限公司

设 计 师：张起铭

参与设计：张在极

项目地点：福建福州

主要材料：石材、墙纸、皮革、木饰面等

◇ 设计理念 ◇

SOHO代表了一种更为自由、开放、弹性的工作方式。网络时代的到来，使得SOHO成为年轻人追逐的一种时髦居住与工作结合的生活方式。在这灵动富有想象的空间里，注重物料质感和设计手法精致表达，赋予它时尚高雅的气质。在陈设艺术的体现上表达了主人对钟表的喜好，对于时光的收藏。空间基调内敛低调就像红酒的醇厚，给人带来细致独享的惬意生活，同时充满国际化的都市情怀。

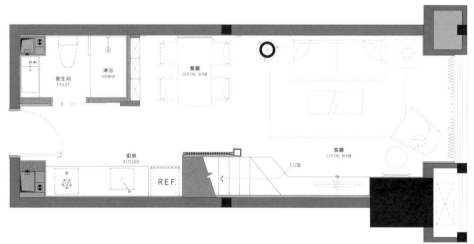

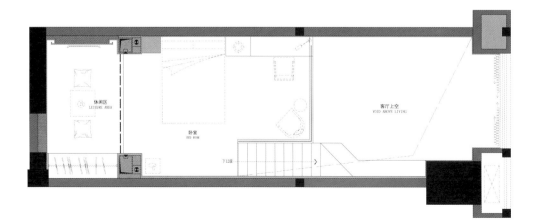

FADDISH CITY
时尚都市

◇ DESIGN CONCEPT ◇

People have their own fashion and lifestyles in mind. The young generation advocate the living atmosphere with fashionable and modern sense. Life is filled with colors, which are full of city flavors like the neon passing through the cityscape and the city. Bright yellow is used on the soft decoration in visual sense. The whole house presents a unique understanding on fashion via the warm and lively colors. While the expression of space is a simple, free and casual living attitude!

项目名称：南屿翡丽湾样板间

设计公司：深圳张起铭室内设计有限公司

设 计 师：张起铭

参与设计：李茜、张在爱

项目地点：福建福州

项目面积：47平方米

主要材料：石材、墙布、黑钢、木饰面等

◇ 设计理念 ◇

每个人的心中，都有自己的时尚和自己的生活方式。年轻一族都崇尚于极具时尚感和现代感的居住味道，生活里充满色彩，这种色彩充满都市情怀就好像穿梭于都市风景及都市的霓虹。软装设计在感观上用了鲜亮的黄色。让整个居室以温暖活泼的色彩呈现出别具一格时尚的见解。然而空间的表达是一种简约、自由和随性的生活态度！

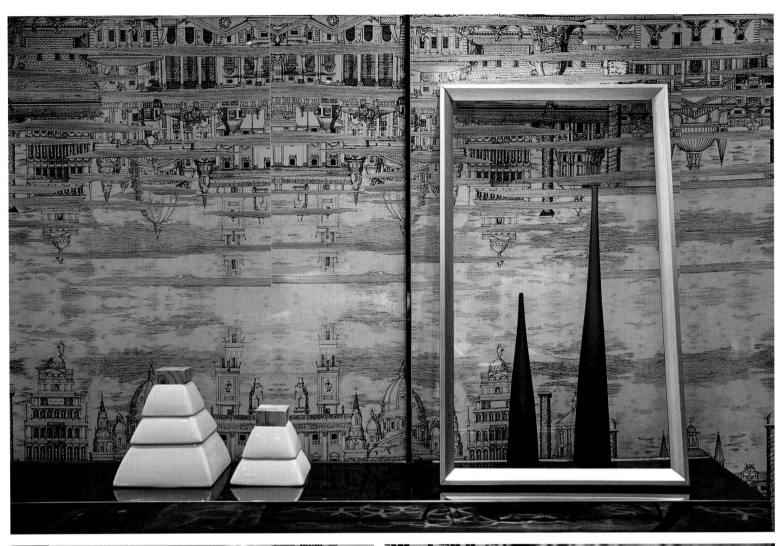

ARTFUL LIFE
艺术生活

◇ DESIGN CONCEPT ◇

This case is combined with modern design techniques, so that the whole space is arranged reasonably. Sliding doors are skillfully used to connect each space where there is tranquility in dynamic. The generatrix at home is no longer simplex, so that the art of design is melt into the family life silently.

The collocation of rose gold steel mirror and leather and wood in dark color is adopted in the living room, which is delicate, gorgeous and low-key without losing fashion. The joining of gold mitigates the steady atmosphere and adds a sense of vitality. Standing here, you will feel not only the appearance of home, but also the soul, warmth and peace. The original layout of the dinning room is abandoned, while semi-connected mode is used to connect the dinning room and the kitchen. Two sliding doors form a rectangular-ambulatory-plane perfectly, making cooking no longer a single action of one person, but a warm interaction of a family. The decorations on the symmetrical display shelf are proper and ordered. The steady color of the leather manifests the owner's unique pursuit and yearning to the quality of life further.

项目名称：顺德天悦湾花园

设计公司：柏舍设计（柏舍励创专属机构）

项目地点：广东佛山

项目面积：235平方米

主要材料：皮革、工艺玻璃、英伦玉、玫瑰金镜钢等

◇ 设计理念 ◇

此案结合现代设计的手法，将整个空间进行合理安排，运用趟门巧妙地连通各个空间，动中有静，家庭动线不再单一，设计的艺术悄然融入家庭生活中。

客厅采用了玫瑰金镜钢搭配深色的皮革和木料，精致而华丽，低调中不失时尚。金色的介入，缓和了沉稳的气息，多了一丝灵动。置身其中，给你的不仅仅是一个家的躯体，还有家的灵魂，家的温馨与祥和。餐厅抛弃了固有的布局，运用半连通的方式，将餐厅和厨房有机地联系在一起。两扇趟门完美地形成回形动线，让下厨不再是一个人的独舞，而是一家人的温馨互动。对称的展示层架，饰物恰到好处，一切井然有序。沉稳的皮革颜色，进一步彰显了主人对生活质感的独特追求和向往。

After the busy work, you could rest in the study to enhance your literacy and think about life. The designer shows his ingenuity on the relation among the study, living room and balcony, where the scenery is blended into the decorations to create the maximal possibility in the limited spaces with the moving of steps, which feasts people's eyes. The light gold color is mellow and warm with the scene of harvest and the feature of impressionism which is deep and restrained, enjoying the tranquility and leisure of the successful life.

工作繁忙之余，小憩书屋，提升自己的素养，感悟人生点滴。还有书房、客厅、阳台三者之间的联系，设计师可谓是匠心独运，将景色融入到装饰当中去，以在有限的空间内，创造出最大的可能性，移步换景，足以让人赏心悦目。淡淡的金色，醇厚而温暖，带着丰收的景象和浓郁而又有所克制的印象主义的特征，似乎是阅尽繁华，享受着成功生活的恬淡与怡然。

The bedrooms are full of layers without any excessive decorations, which pursues for the experience of great quality instead of luxury, collocated with the effect of lights, creating a sense of sunshine through the refraction of the steel mirror.

卧室层次感十足,没有过分的修饰,追求的不是奢华,而是高尚品质的体验。搭配灯光的效果,镜钢的折射营造出阳光的感觉。

METROPOLITAN
精品大都会

◇ DESIGN CONCEPT ◇

Baohua Jakaranda Garden Show Flat Trilogy Type G is in metropolis style which assembles all the thoughts on boutique goods. The appearance of this style is the thought and embodiment of human civilization, which is also the inner reflection of metropolis people. Everyone is searching for his or her own character.

Thus it is inclusive and unique, which blends the order and disorder together. In the space, leather combines with the chrome plated metal with decorative lines in industrial sense, while khaki and white are used as the background color. The embellishment of some colors in warm tone makes it different from the cold feeling in modernism. The brass trumpet changes into the brass cookware in the kitchen, which is a classical symbol. The attitude towards culture and various elements is: all-inclusion.

项目名称：宝华·紫薇花园G户型
设 计 师：连自成
参与设计：金李江、耿小丽
项目地点：上海
项目面积：270平方米
主要材料：胡桃木、毛面砖、铜艺栏杆、大理石、陶瓷、铜镀铬等

◇ 设计理念 ◇

宝华·紫薇花园样板房三部曲之G户型，是荟萃了所有对于精品思考的大都会风格，这样的风格出现是对人类文明的思考和体现，是居住于都市中的人们，内心的反应。每个人都在寻找一个适合自己的角色。

于是它是包容的，是洋气的，是将秩序感和混乱感融合在一起的。空间里，皮革材质与精密平整的镀烙金属相结合，装饰线条呈工业设计感，卡其色白色作为背景色。穿插其中一些暖色系的点缀，有别于现代主义冷冽的感觉。黄铜喇叭的印象，转化成了厨卫里的铜制锅具，经典的象征。对待文化、对待各种不同元素的态度：兼容并包。

The home decoration is vogue, neat and upscale which is suitable for those who pursue for delicate life forever. A home could exclude the unique charm at any time with fashion shoulder by shoulder. The broad room, vogue and elegant modern furniture from Fendi reveal a low-key luxury. The modern kitchen with high-tech and high quality is the proper attitude of modern and fashion in this metropolis. This lifestyle pursues for simplicity yet remains the gorgeousness. Modern metropolis culture is liberated from modern simple design, while the concise concept that Bauhaus emphasizes means that decoration is a kind of evil. While metropolis spirit is the sarcasm on simplicity, which uses decorative arts after redefined to emphasize that luxury is innocent. This is the reflection of the urbanites' confidence as well as the yearning towards human civilization.

家饰时髦、整洁和高端，它适合永远追求精致生活的人群，家能够时时刻刻散发出独特的魅力，且永远和时尚齐头并进。开阔的房间，时髦且雅致的Fendi现代家具，流露出一种低调的奢华，高科技高品质的现代化厨房，是这个城市应有的摩登时尚的态度。这种生活方式追求简洁但是同时保持着华丽。现代的大都会文化是从现代简约设计中解放出来，包豪斯强调的简约概念表示装饰就是一种罪恶。而大都会精神即对于简约的讽刺，用重新定义的装饰艺术去强调了奢侈无罪。这是都市人的自信反应也是对人类文明的向往。

KEY C WITH FASHION
时尚C调

◆ DESIGN CONCEPT ◆

It is said that a space is like a person. This show flat occupies the exquisite and rich connotation under the silent and cold appearance, which expresses the attitude of "not flaunt optionally or sequacious egoistically" to life. Designers use the technique of the fashionable gray tone to reflect the pursuit on living quality of some middle-class people in metropolis life as well as their unique world view, which not only stresses on details but also the practicability of the functions. A small round table is allocated in the dinning room to embellish the uneasiness of the square space, while the wardrobe meets the needs of storage and setting of goods. A tea room with meditation function is elaborately built to inject several Zen flavor into this modern space, noticing people not to forget to calm down and deposit out of the busy daily life. The temperament of the space is conveyed through different materials and soft decorations, attempting to make each area bring a different visual experience.

项目名称：卢卡小镇联排别墅样板房

设计公司：上海无相室内设计工程有限公司

设 计 师：王兵、王建

软装设计：李倩

项目地点：陕西西安

项目面积：190平方米

摄 影 师：张静

主要材料：意大利木纹石、紫檀木皮、香槟金不锈钢、石材马赛克等

◆ 设计理念 ◆

都说空间如人，此样板房在沉默冷峻的外表下内涵细腻丰富，表达对生活"不随意张扬，不任性盲从"的态度。设计师以时尚灰调的手法来体现都市生活中对生活品质的追求，不仅注重细节，同时强调功能的实用性。餐厅配置小型圆桌，修饰方形空间的局促感，到顶立柜满足收纳、置物需求。刻意打造品茶冥想室，给现代空间注入些许禅意，提醒在忙碌的日常生活之余不忘静心、沉淀。空间的气质则通过不同材质、软饰来传递，意在让每个区域都能给人带来不同的视觉体验。

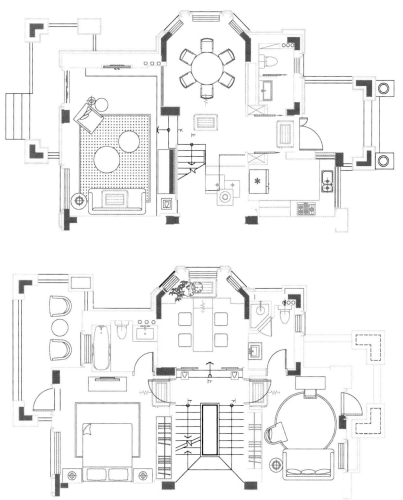

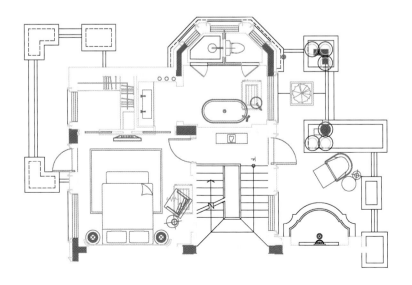

Large area of red sandalwood is used as the surface material of the decorations and furniture, while wallpaper, stainless steel with champagne gold surface and other modern materials are used properly to balance the visual focus, collocated with abstract art paintings, creating a wild individual space where people could taste its connotation slowly. The main placing in the space gives a lengthways feeling of expansion, however, the apricot sofa with three kind of cushions of cotton, wool and silk echoes with the carpet in geometrical patterns, harmonizing the overall feeling in the space. Besides emphasizing on the texture of soft decorations, designers use effective lighting design to bring the furnishings into fashion from another aspect. The shadow makes the metropolis life no longer the flashy fashion, but the comfort with high quality.

此案以大面积的的紫檀木作为装饰及家具的表面材料，适当运用墙纸、香槟金不锈钢等多种现代材料平衡视觉焦点，搭配抽象艺术画作，构造一个狂野不羁却可以让人细细品味其内涵的个性空间。空间的主体摆放，很有纵向的拓展感觉，客厅的杏色沙发摆着棉、毛、丝质的三种抱枕，呼应几何纹图案的地毯，协调了空间的整体感觉。除了重视软饰品的质感享受，有效的灯光设计将陈设物品从另一个角度带入时尚的步调，渲染的光影让都市的生活不再是浮华的时尚，而是有品质的舒适。

AMERICAN STYLE / 美式风格
NEO-CHINESE STYLE / 新中式风格
MODERN LUXURIOUS STYLE / 现代奢华风格
SOUTHEAST ASIAN STYLE / 东南亚风格

SOUTHEAST ASIAN STYLE
东南亚风格

A VACATION IN SOUTHEAST ASIA
度假东南亚

DESIGN CONCEPT

Southeast Asian style is famous for its bold color collocation, which is open yet connoted, charming yet mysterious, calm yet passionate in the gorgeous "hot dance". In this case, we used light tones and adopted large amount of natural woods and fresh wallpapers to give a soft and gentle feeling. The living room and dinning room were based on the generousness and elegance, while the background of mandragora was collocated with large area of light wood veneer modeling, exuding a strong natural sense.

项目名称：长泰淀湖观园二期
设计公司：上海牧笛室内设计工程有限公司
设 计 师：毛明镜、昌影
项目地点：江苏昆山
项目面积：215平方米

设计理念

东南亚风格素以大胆的配色著称，在绚丽的色彩"热舞"中，舒张而有含蓄，妩媚又有神秘，平和却有激情。在本案设计中我们以浅色系为主，大量采用了纯天然的木料和自然清新的壁纸给人轻柔的感觉。客餐厅以大气优雅为主，曼陀罗花背景配以大面积的浅色木饰面造型，散发着浓烈的自然气息。

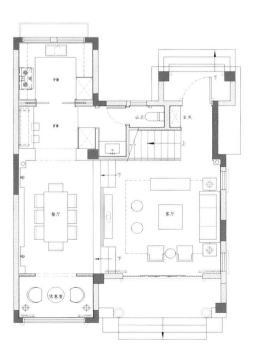
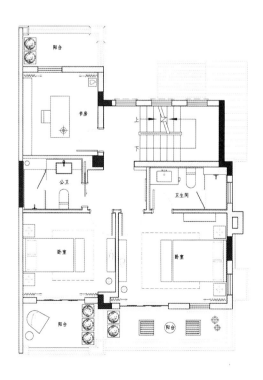

Wood grating was used in master bedroom to divide the spaces, in order to meet the functional demands and use spaces reasonablly instead of all the complex and decorations. The purdah created the graceful and harmonious magnificence through the bold colors and delicate collocations. Some blue and yellow tones were added in other spaces to make the space not appear so calm. The furniture were based on natural vine and wood tone, especially brown and other dark colors, forming the pristine sense of soil on visual feeling, added with the embellishment of fabrics, making the atmosphere active without monotonousness.

主卧室运用木格栅区分空间满足功能需求、合理利用空间，代替一切繁杂与装饰，床帏处的帐幕、香艳而浓烈的色彩，以大胆的配色和精巧的搭配创造出华美和谐的绚丽感。在其他空间中加入一些蓝色和浅黄色调，这样处理过后空间看上去不过于沉稳。家具色泽以原藤、原木的色调为主，大多为褐色等深色系，在视觉感受上有泥土的质朴，加上布艺的点缀搭配，非但不会显得单调，反而会使气氛相当活跃。

In the interior design of southeast Asian style, leisure, freedom and faintness all appeared in the wispy curtain. The silk purdah hanging on the bedstead created a dreamy sky for the sweet dream in the afternoon. The wind blowed through the balcony, with plants, fragrances, water and flowers, forming the implied romance.

东南亚风格的室内设计中，闲适、安逸、隐约都在飘渺的纱幔中尽显出来。床架上挂着的丝质纱幔，给午后的甜梦营造一方隐约的梦幻天空。阳台上微风吹拂、枝枝蔓蔓、阵阵熏香、水与花交织成的浪漫成了潜在的一缕脉动。

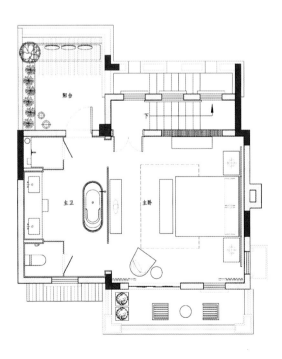
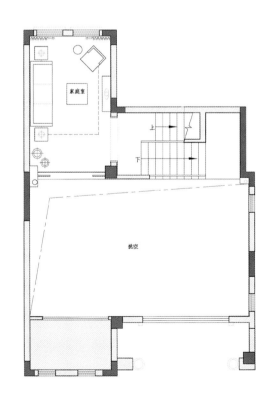

LOVE IN SOUTHEAST ASIA
情定东南亚

◇ DESIGN CONCEPT ◇

Rattan furniture is the feature of southeast Asia, the casual and asymmetric structure of which gives the inhabitants a more relaxed leisure. Natural broad leaved plants and bright flowers reflect the state of integration. In a room full of tropical flavor, the white wall is collocated with the carved hollow patterns with Southeastern feature with gold lines, appearing the wealth and harmony. The overall space is dominated by dark tone such as dark brown, black and gold, which is calm, mysterious and generous. At the same time, the features of the Southeastern furniture are remained through the collocation of different materials, producing more colorful changes. For example, there is an elegant sofa with dark wood frame when being looked from one side, delicate and exquisite. Through the shining of gentle light in warm tone, a warm and comfortable atmosphere is pervaded in the overall concise and broad space. The appropriate elaborate arrangement brings more luxury and nobility into the space.

项目名称：世茂苏州御珑墅
设计公司：上海李峻设计咨询有限公司
设 计 师：李峻、武帅
项目地点：江苏苏州
项目面积：350平方米
摄 影 师：美宿摄影
主要材料：木饰面、大理石、硬包等

◇ 设计理念 ◇

藤艺家具是东南亚风格的代表特征，随性散布不对称的摆设格局给居住者更轻松的休闲。自然的阔叶植物，鲜艳的花卉，体现物我相融的境界。在一间充满热带风情的居室中，用洁白的墙壁搭配东南亚特色的雕刻镂空图案，并描以金色，富贵祥和。整体空间以浓郁的深色系为主，如深棕色、黑色、金色等，沉稳、神秘、大气。同时，通过不同的材质搭配令东南亚家具设计在保留了自身的特色之余，产生更加丰富多彩的变化。如一张设计幽雅的沙发床，从侧面看，洁白的床垫搭配深色的木框，别致精巧。透过柔和的暖色灯光照射，整个简约开阔的空间里，处处弥漫着温馨舒适的氛围。不多不少的精心装置，带给空间更大的豪华与高贵。

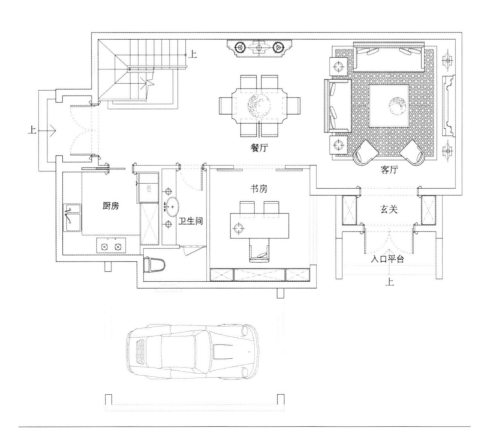

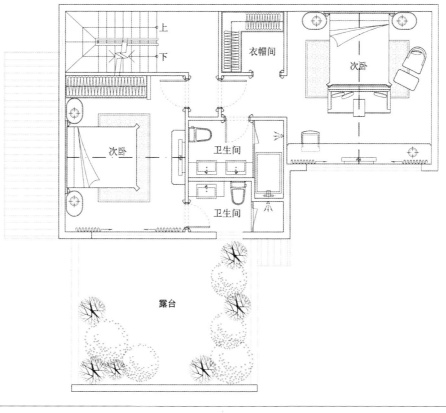

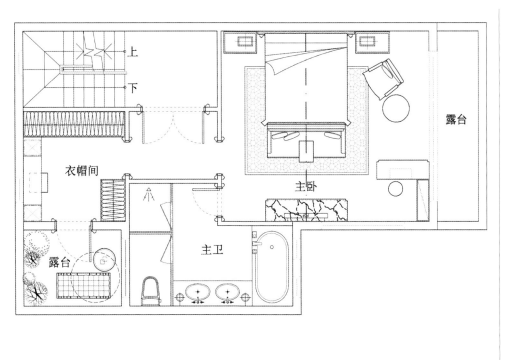

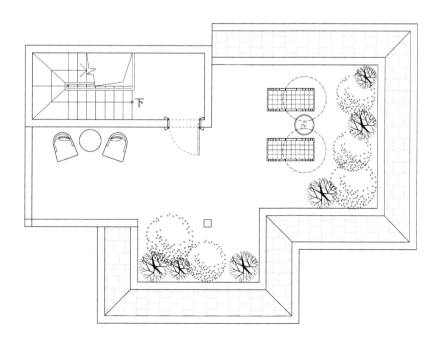

A SOUTH ASIAN RESIDENCE WITH ZEN FLAVOR
南亚禅境

◆ DESIGN CONCEPT ◆

The furniture full of South Asian flavor matched with colorful decorations, while the colors in different depth added more vitality into the original wood color which was heavy, embodying the ebullient and unsophisticated culture and custom in South Asia. A holiday atmosphere just blew here, and the air seemed like the specific damp and hot sense of South Asia with the natural fragrance of woods, flowers and agilawood. You could enjoy the relaxed and leisure experience of holiday without going abroad. Far away from the disturbance of the hustle and bustle in cities, this scene seemed like untruthful, which made you feel as if you were standing in the fairyland yet there was a rich Zen flavor. Designers echoed each theme and posture of different spaces through using design languages such as colors and materials. In this series of works, you could realize that each theme is a surprise.

项目名称：无锡灵山小镇•拈花湾系列作品之样板区

设计公司：禾易HYEE DESIGN（原HKGGROUP）

设 计 师：陆嵘

参与设计：李怡、卜兆玲、王玉洁

项目地点：江苏无锡

项目面积：198平方米

◆ 设计理念 ◆

充满南亚风情的家具搭配着色彩缤纷的装饰品，深深浅浅不同层次的颜色给原本沉重的木色增添更多活力，体现出南亚热情淳朴的民风民俗。度假气息铺面而来，吸入的空气也像是南亚特有的湿热味道，混着自然的木香、花香、沉香，不出国门也能享受这轻松悠哉的假日体验。远离了城市喧嚣的干扰，这画面似乎有些不真实，仿佛置身于世外桃源，但它们却透着浓浓的禅味儿。设计师通过运用颜色、材质等设计语言，呼应着各自的主题、舒展着不同的姿态。在这一系列的作品中，你可以读出每个主题都是一次惊喜。

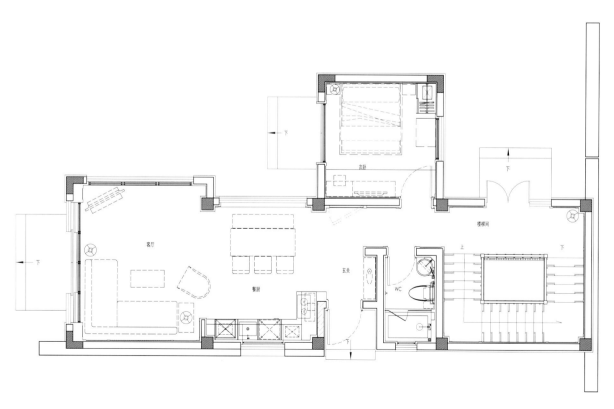

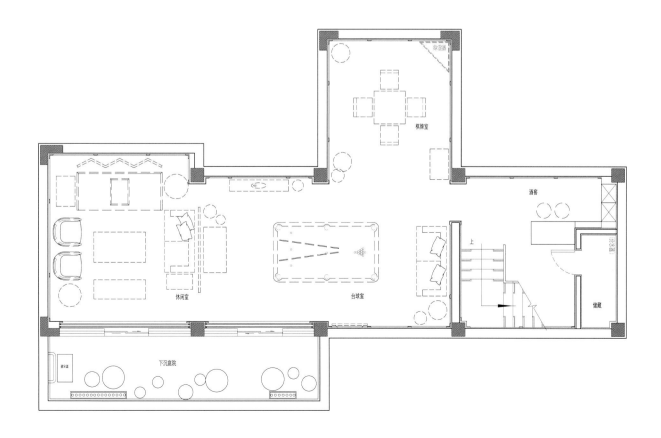

A PURE REGION
净域

◆ DESIGN CONCEPT ◆

Located in the east coast of Huangpu River, this house type is a single villa with four floors, oriented as Southeast Asian style.

The high-ceiling design at the entrance foyer at the first floor adds the sense of vertical extension in the space, while the French windows bring the outdoor scenery into the room, forming a wonderful opposite scenery with the stean and sculpture of Buddha head. The double-floor high-ceilinged living room manifests the generousness and elegance of the space. There is a leisure studying area through the half-enclosured wooden partition, which is the outspread space of the living room. Colorful sofa and cushions are collocated in the living room to avoid the sense of repression and break the feeling of heaviness on visual sense.

项目名称：新浦江城十号院

设计公司：One House壹舍室内设计（上海）有限公司

设 计 师：方磊

参与设计：李文婷、李丽娜

项目地点：上海

项目面积：730平方米

主要材料：米白洞石、实木、布艺硬包、拉丝古铜、藤编等

◆ 设计理念 ◆

本案地处黄浦江的东岸。此户型是四层的单体别墅，定位为东南亚风情。

一楼入口玄关的挑空设计，增加空间的纵向延伸感，横向上落地窗将室外景观引入室内，与摆放的陶罐和佛首构成一处巧妙的对景。客厅的双层挑高，彰显空间的大气与优雅。透过半围合的木百叶隔断是个休闲学习区，是客厅的延展空间。客厅中搭配色彩艳丽的沙发和靠垫抱枕，避免了空间的压抑，冲破视觉的沉闷。

The dinning area is adjacent to the living room, which is connected with the outdoor area. You could have a rest on the terrace outside and enjoy the beautiful sceneries. Featured lounge and accessories with southeastern style are collocated in the interior space to create a relaxed and leisure atmosphere. A half-transparent partition divides the master bedroom and sitting room, forming a comparison with the materials on the wall. Dark furniture, silk texture and linen fabric are matched as well as the change of shadows to make the room exude a light warmth and Zen flavor.

紧邻客厅的是餐厅区域，餐厅与户外相通，在闲暇之余可以在户外平台休憩，享受户外的美景。家庭室内搭配特色卧榻及具有东南亚风格的配饰，来营造出轻松、休闲的空间气氛。主卧室与起居室由半通透的隔断将两个空间分开，和墙面材料形成虚实的对比。搭配深色的家具，丝绸质感和亚麻的布料，结合光影的变化，使居室散发着淡淡的温馨与悠悠的禅韵。

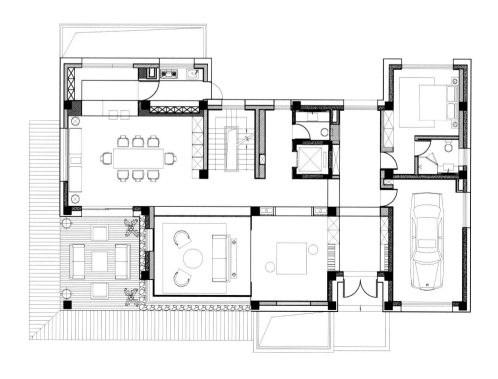

Wood color is dominated in the SPA area, added with concise and fluent lines, making this area a tranquil, warm and comfortable leisure space. It is natural, pristine and dark-colored with tropical flavor, which is a design combined with features of southeast Asian islands and delicate cultural taste, where large amount of wood and other natural materials are used. The furniture in dark color and the change of lights embody the steadiness and luxury sense, conveying the design concept which is leisure and free yet luxurious.

SPA区以木色为主色调,加上简约流畅的线条分割,使SPA区成为一个静谧、温暖、舒适的休闲空间。自然、朴实、色彩深沉,有热带风情,是东南亚民族岛屿的特色及精致文化品味相结合的设计,广泛地运用木材和其他的天然原材料。深木色的家居,灯光的变化体现了稳重及豪华感。传达出了既可以悠闲自在,也可以奢华的设计观。

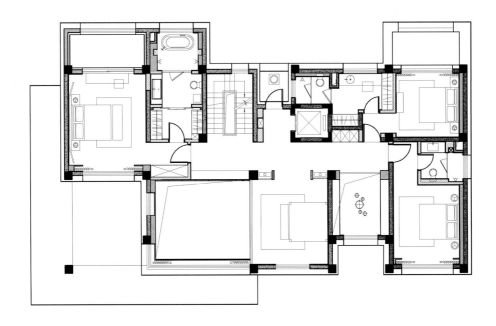

The southeast Asian feature of advocating nature is expressed through stone, wood and rattan in the space to bring a heaviness on visual sense. Natural materials are adopted in the design to present all the details in the best way. The overall space implies a comfortable atmosphere no matter it is eastern or exotic. Here, you could keep away from the hustle and bustle and return to nature quietly.

空间中将东南亚崇尚自然的特性通过石材、实木及藤条来表达，给视觉带来厚重感。设计中采用的天然材料，让所有细节得以最好呈现。整个空间蕴藏着一股难以言喻的舒适气息，或东方，或异域。远离喧嚣，回归自然，置身于谧。

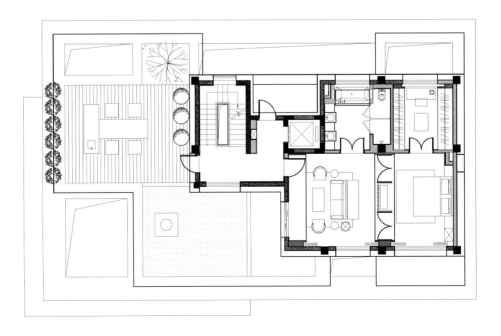

A CHANSON IN GREEN
绿茵香颂

◇ DESIGN CONCEPT ◇

Themed at green, this design scheme created a natural atmosphere where human mind could be relaxed and returned sufficiently. Olive green was used as the basic tone, combined with earth colors, presenting a calm and lively living atmosphere. The exuberant green plants were the feast for eyes, the particular breath of life could create the atmosphere of nature to satisfy people's yearning for green. The jungle scenery that presented in the space brought people into the theme directly, where there were flowing water and fragrance, making people feel warm and relaxed as well as take the tests on visual, auditory and tactile sense to achieve the tranquility on both body and mind as well as forget the hustle and bustle. The implication of returning to nature throughout the whole space aimed at creating a tranquil atmosphere separated from the world. In such a space themed at green, imagine that there is a man with a cup of tea and a good book at the sunsets, how intoxicating it will be.

项目名称：苏州御江南别墅样板房
设计公司：苏州西木莲软装工程有限公司
硬装设计：钱玉民
软装设计：栗微
项目地点：江苏苏州
项目面积：400平方米
摄影师：杨森
主要材料：水木饰面、大理石等

◇ 设计理念 ◇

此设计方案以绿色为主题，营造如同置身自然的氛围，让人的心灵充分放松和回归。以橄榄绿为基调，融合大地的色彩，呈现出沉稳又活泼的生活氛围。绿色植物生气盎然，令人赏心悦目，独有的生命气息可以营造出大自然的氛围满足人们对绿色的向往。空间里所呈现的丛林景观将人们直接引入主题，当中流水潺潺，暖香扑鼻，温暖而放松，让人们在短短的逗留当中得视觉、听觉、触觉的洗礼，达到身心的宁静，忘却尘嚣。回归自然的寓意始终贯穿整个空间，旨在营造一种脱离尘世的静谧氛围。在这样一个以绿色为主题的空间里，每当夕阳西下，一个人、一杯红酒、一本好书，想想也醉人。

图书在版编目（CIP）数据

献计献策：样板房设计新法. 下册 / 深圳视界文化传播有限公司编. -- 北京：中国林业出版社，2015.9
　ISBN 978-7-5038-8151-0

Ⅰ. ①献… Ⅱ. ①深… Ⅲ. ①住宅－室内装饰设计－图集 Ⅳ. ①TU241-64

中国版本图书馆CIP数据核字(2015)第221519号

编委会成员名单
策划制作：深圳视界文化传播有限公司（www.dvip-sz.com）
总 策 划：蒙俊伶
校　　对：丁　涵　杨珍琼
翻　　译：曹　鑫
装帧设计：万　晶　潘如清
联系电话：0755-82834960

中国林业出版社　·　建筑家居出版分社
策　　划：纪　亮
责任编辑：纪　亮　王思源

出版：中国林业出版社
（100009 北京西城区德内大街刘海胡同 7 号）
http://lycb.forestry.gov.cn/
电话：（010）8314 3518
发行：中国林业出版社
印刷：深圳市汇亿丰印刷科技有限公司
版次：2015年9月第1版
印次：2015年9月第1次
开本：230mm×300mm，1/16
印张：23
字数：150千字
定价：398.00元（USD 81.00）